Textile Progress

June 2010   Vol 42   No 2

A roadmap on smart textiles

T0186907

Anne Schwarz,
Lieva Van Langenhove,
Philippe Guermonprez
and Denis Deguillemont

The Textile Institute

Taylor & Francis

Taylor & Francis

SUBSCRIPTION INFORMATION

Textile Progress (USPS Permit Number pending), Print ISSN 0040-5167, Online ISSN 1754-2278, Volume 42, 2010.

Textile Progress (www.tandf.co.uk/journals/TTPR) is a peer-reviewed journal published quarterly in March, June, September and December by Taylor & Francis, 4 Park Square, Milton Park, Abingdon, Oxon, OX14 4RN, UK on behalf of The Textile Institute.

Institutional Subscription Rate (print and online): $335/£176/€267
Institutional Subscription Rate (online-only): $319/£168/€254 (plus tax where applicable)
Personal Subscription Rate (print only): $123/£63/€98

Taylor & Francis has a flexible approach to subscriptions enabling us to match individual libraries' requirements. This journal is available via a traditional institutional subscription (either print with free online access, or online-only at a discount) or as part of the Engineering, Computing and Technology subject package or S&T full text package. For more information on our sales packages please visit www.tandf.co.uk/journals/pdf/salesmodelp.pdf.

All current institutional subscriptions include online access for any number of concurrent users across a local area network to the currently available backfile and articles posted online ahead of publication.

Subscriptions purchased at the personal rate are strictly for personal, non-commercial use only. The reselling of personal subscriptions is prohibited. Personal subscriptions must be purchased with a personal cheque or credit card. Proof of personal status may be requested.

Ordering Information: Please contact your local Customer Service Department to take out a subscription to the Journal: **India**: Universal Subscription Agency Pvt. Ltd, 101–102 Community Centre, Malviya Nagar Extn, Post Bag No. 8, Saket, New Delhi 110017. **USA, Canada and Mexico**: Taylor & Francis, 325 Chestnut Street, 8th Floor, Philadelphia, PA 19106, USA. Tel: +1 800 354 1420 or +1 215 625 8900; fax: +1 215 625 8914, email: customerservice@taylorandfrancis.com. **UK and all other territories**: T&F Customer Services, Informa Plc., Sheepen Place, Colchester, Essex, CO3 3LP, UK. Tel: +44 (0)20 7017 5544; fax: +44 (0)20 7017 5198, email: subscriptions@tandf.co.uk.

Dollar rates apply to all subscribers outside Europe. Euro rates apply to all subscribers in Europe, except the UK and the Republic of Ireland where the pound sterling price applies. If you are unsure which rate applies to you please contact Customer Services in the UK. All subscriptions are payable in advance and all rates include postage. Journals are sent by air to the USA, Canada, Mexico, India, Japan and Australasia. Subscriptions are entered on an annual basis, i.e. January to December. Payment may be made by sterling cheque, dollar cheque, euro cheque, international money order, National Giro or credit cards (Amex, Visa and Mastercard).

Back Issues: Taylor & Francis retains a three year back issue stock of journals. Older volumes are held by our official stockists to whom all orders and enquiries should be addressed:
Periodicals Service Company, 11 Main Street, Germantown, NY 12526, USA. Tel: +1 518 537 4700; fax: +1 518 537 5899; email: psc@periodicals.com.

The 2010 US Institutional subscription price is $335. Periodical postage paid at Jamaica, NY and additional mailing offices. **US Postmaster:** Send address changes to TTPR, c/o Odyssey Press, Inc., PO Box 7307, Gonic NH 03839, Address Service Requested.

Subscription records are maintained at Taylor & Francis Group, 4 Park Square, Milton Park, Abingdon, OX14 4RN, United Kingdom.

For more information on Taylor & Francis' journal publishing programme, please visit our website: www.tandf.co.uk/journals.

CONTENTS

Textile Progress
Vol. 42, No. 2, June 2010, 99–180

A roadmap on smart textiles

Anne Schwarz[a]*, Lieva Van Langenhove[a], Philippe Guermonprez[b]
and Denis Deguillemont[b]

[a]*Department of Textiles, Ghent University, Zwijnaarde, Belgium;* [b]*Institut Francais du
Textile et de l'Habillement, Villeneuve d'Ascq, France*

(*Received 3 November 2008; final version received 4 November 2009*)

Though industrial exploitation of smart textile systems is still in its infancy, the technological implementation is increasing. This is the result of substantial research and development investments directed towards this emerging field. In order to stimulate the progress in smart textiles, emerging developments need to be identified and selectively strengthened. Hence, this issue reports on a three-dimensional roadmap on smart textiles. It aims at contributing to set future actions in research, education and technology development. Research activities and technological developments are mapped, barriers and drivers of technological, strategic and societal and economical origins are identified. Finally, recommendations are phrased on how to overcome barriers and to progress in the field of smart textiles.

Keywords: roadmap; smart textiles

This issue represents a roadmap report on smart textiles prepared in the framework of the European project Clevertex, which ended in spring 2008.

In general, the Clevertex project aimed at developing a master plan and framework for future actions in research, education and technology transfer in the field of smart textile materials in Europe for transforming the industry into a dynamic, innovative, knowledge-driven competitive and sustainable sector by 2015.

Moreover, the objectives were to map the possible future technological developments in the smart textile sector from a socioeconomic (non-technological) and technological perspectives, to prioritise these possible developments in the actual socioeconomic environment (technology foresight) and to identify needs, breakthroughs and bottlenecks in order to answer these developments. On a "macro-level", the study contributes to the "new economics of science" where the allocation of resources is based on strategic considerations. The long-term foresight study aims at identifying research and scientific priority areas.

A well coordinated and focussed approach was needed, combining all potential contributors to scientific and technological development into a converging positive direction. This required a focussed effort in research and development, transfer of technology, education and training based on cooperation between industry, government and research centres. To do this in an efficient and effective way, four partners, being Institut Français du Textile et de l'Habillement (IFTH), Scientific and technical centre of the Belgian Textile industry (Centexbel), Technical University of Lodz/Poland and Ghent University, were working

---

*Corresponding author. Email: Anne.Schwarz@ugent.be

ISSN 0040-5167 print/ISSN 1754-2278 online
DOI: 10.1080/00405160903465220
http://www.informaworld.com

together. Five industrial partners from Europe, namely Nota (Greece), Alcatel (France), Smartex (Italy), Andixen (France) and Bonfort (Belgium), supported the research institutions. Above that, the European textile and clothing association (EURATEX) assisted the consortium with their skills and expertise in performing foresight studies in the area of information technologies for the textiles and clothing industry.

1. Introduction

The textile and clothing industry in Europe has already faced a structural change for decades. The relocation of production facilities to low-wage countries and increasingly fierce global competition characterise this situation. Nevertheless, the textile and clothing industry still presents one major industrial sector and is therefore still of major importance for the social and economic welfare in Europe.

The three main activities in this industry are clothing (46% of the total production), home and interior textiles (32% of the total production) and technical textiles (22% of the total production) as shown in Figure 1 [1].

To preserve its leading role in the global market, the focus in Europe is increasingly directed on creative fashion design, quality, innovation and re-organisation in manufacturing as well as vertical integration and consumer brand building. The removal of import quotas in 2005 for textile products due to the liberalisation of industrial nations' markets enhances even more the necessity for the industry to differentiate itself.

Differentiation is possible by specialising on the products' quality and functionality as well as on the flexibility and quick response of the services rather than on the price alone. The necessary precondition to achieve this goal is the deployment of the latest research results and active research in smart textiles, a field ranging from materials science, textile engineering, chemistry and electronics to informatics by a highly qualified workforce (Figure 2). It will further require applied research at long-, mid- and short-term levels to transfer the technology to companies which must adapt the knowledge and consequently bring new and value-added products to the market.

The trend towards differentiation is additionally supported by the constant decline of consumer income spent on clothing and textiles for home decoration. Consumers' preferences gained in the field of healthcare, well-being and sports activities as well as performance and protective equipment, which resulted in increasing profits of clothing sectors like outdoor and sportswear, protective clothing and work wear. During past years many new enterprises specialising in niche markets in the field of innovative and functionalised textiles

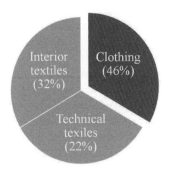

Figure 1.   Main activities of the textile industry; clothing represents the largest branch, followed by the interior textile sector. The technical textile market has the lowest production (data taken from [1]).

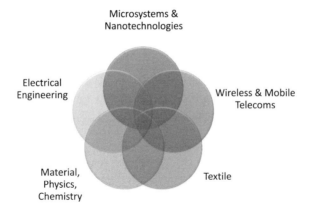

Figure 2. Converging different disciplines to develop smart textiles.

and clothing have appeared. They achieved competitiveness by exploiting the opportunities of new and innovative materials and non-conventional textile applications [2].

This document aims at contributing to the economic and social progress of the textile and clothing sector, by identifying areas of research in smart textiles that can have a direct and positive impact on the health and living conditions of the society.

Against this background, the present issue aims at providing a plan to bring smart textiles to the forefront of value-added textile products. The document provides an opportunity to set a direction for future research and technological development and to determine the priorities of research and product development plans in research organisations, industry and policy decision makers.

1.1. Vision

We live in a knowledge-driven society that faces an increasing impact of science and technology on all aspects of life through products and services, and consumer needs. The field of smart textiles is not yet a discrete area; it is more an interdisciplinary subject incorporating science, technology, design and human sciences, and its future lies in the potential of combining different technologies. The convergence of textiles and electronics can be exemplarily pointed out for the development of a smart material, which is capable of accomplishing a wide spectrum of functions found in rigid and inflexible electronic products today. Smart textiles could serve as a mean to increase the well-being of society and they might lead to important savings on the health budget. Moreover, they could increase consumption, as they are not only a high-value product but also an immaterial concept that satisfies the consumer needs and demands such as creativity and emotional fulfilment.

In other words the successful introduction and application of smart textiles will allow the production of new generations of innovative and high value-added textile products. Smart textiles will yet be affordable, possible to adapt and benefit a sustainable and competitive textile industry development.

1.2. Methodology for roadmapping

The following paragraphs give an insight into the common roadmap methodology applied by various individual companies and whole research and business sectors. The structure of this roadmap is described as follows.

1.2.1. What is a roadmap?

In general, a "road map" is a layout of paths and routes for travellers to show the direction and proximity in a geographical area.

Recently, the term "roadmap", has become a popular synonym for a strategy or a project plan. It has been used in industry, government and the scientific world to portray the structural relationships between science, technology and applications by roughly planning different steps to follow over a perennial period of time. Thus, they are used as decision aids to improve coordination of activities and resources in increasingly complex and uncertain environments.

So far, an overall valid definition of a "roadmap" does not exist. It was first coined by Motorola in the 1970s and it was Robert Galvin, former chairman of the board of directors for Motorola, who offered a first definition in the late 1990s [3]:

> A "roadmap" is an extended look at the future of a chosen field of inquiry composed from the collective knowledge and imagination of the brightest drivers of change in that field. "Roadmaps" communicate visions, attract resources from business and government, stimulate investigations, and monitor progress. They become the inventory of possibilities for a particular field.

Therefore, in the broadest sense, the major benefits of a roadmap are to

- help to develop consensus among decision makers about a set of needs (technological, economic, societal, strategic);
- help to forecast developments in targeted areas;
- present a framework to help plan and coordinate developments at any level within an organisation, throughout an entire industry and even across industry or national boundaries.

Coming back to the classical roadmap, the map of London's underground is shown in Figure 3. It can be seen that it consists of two dimensions, a horizontal and a vertical line, to portray the crossings and directions of the different metro lines [4]. Likewise, a roadmap also consists of two dimensions, a spatial and a temporal dimensions (Figure 4). The spatial dimension reflects the relationship among the disciplines at a given point of time. The time dimension accounts for the evolution of the same capabilities [5].

Finally, it should be noted that the true extent of the benefits has yet to be proven. However, it is already clear that it is an effective tool to structure industry–government research programmes and to facilitate collaboration within industries and among companies and thus it is increasingly being used.

1.2.2. How is the roadmap built up?

The methodology used for this roadmap followed the procedures typically used for other roadmaps that have been developed. Basically, four main stages of the roadmapping process were followed by asking the following:

1.2.2.1. Where are we now? The first step was to identify the present situation of smart textiles as well as the socioeconomic situation of the society. In order to identify the current research and product landscape of smart textiles, this document includes a chapter on the state-of-the-art of smart textiles.

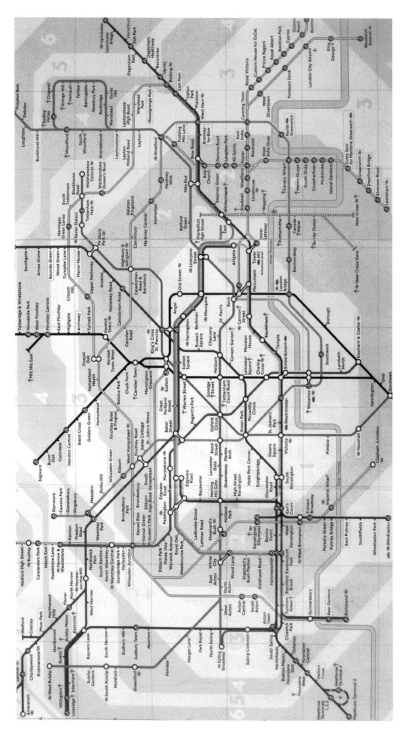

Figure 3.   A classical roadmap: map of the underground in London. Picture by author.

Capacities

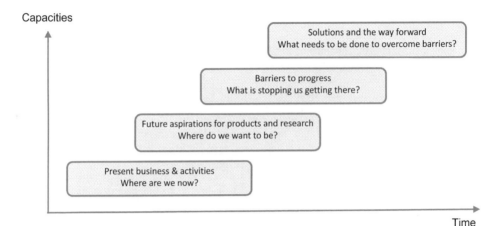

Figure 4. The two dimensions of a roadmap – time as the horizontal line and capacities on the vertical line.

1.2.2.2. Where do we want to be? In the second step, a vision for the future to move on was provided by means of a Delphi study, supported by commonly published and well-known forecasting studies (for instance VDC and BCC). Above that, scenarios describing everyday situations in the years 2020 and 2030 contributed to envisaging our future lives with smart textiles.

1.2.2.3. What is stopping us from getting there? The third stage was to determine possible barriers towards our vision. The barriers were split up in the following three dimensions:

1. Strategic barriers.
2. Societal and economic barriers.
3. Technological barriers.

Barriers for the successful implementation of smart textiles in industry and in our daily life are described in a general manner. Above that, specific market studies were carried out to identify the barriers for five defined markets and are presented subsequently.

1.2.2.4. What has to be done to overcome these barriers? To come over these barriers and to filter out the most promising ways to follow, decisions and proposals need to be made. Finally, recommendations are formulated and actions to follow are suggested.

The different steps of the roadmap, as illustrated in Figure 5, are closely linked to and built up on each other. In the following chapters of this report, we describe and examine these steps in detail.

2. Defining a smart textile system

2.1. Definition of smart textiles

In the first phase, a study on smart textiles is reduced to a study on smart materials. In the second phase, it is to be considered in which way these smart materials can be processed

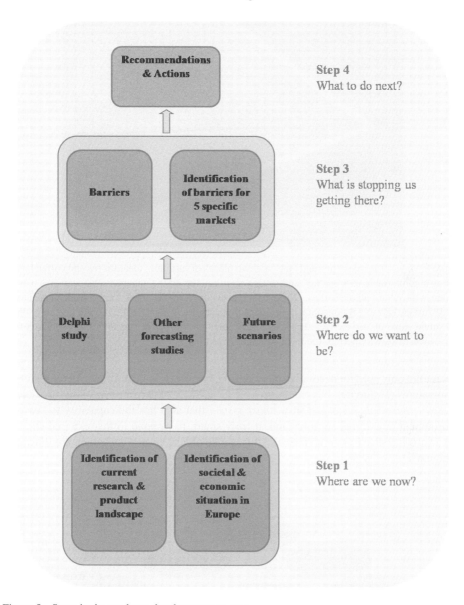

Figure 5. Steps in the roadmap development process.

into a textile material. For that reason we first consider the definition of smart materials in general.

The most commonly accepted definition is that *intelligent materials* and *systems* are capable to sense and respond to their surrounding environment in a predictable and useful manner [6]. *Smart materials* arise from research in many different fields. One example are thermochromic materials, changing their colour in response to changes in temperature. They can be made as semiconductor compounds, from liquid crystals or using metal compounds. The change in colour happens at a determined temperature, which can be varied by doping the material. As they are used to make paints, inks or are mixed to moulding or casting materials for different applications, their application is manifold. In textiles, thermochromic

Figure 6. Interior textiles "tic" and "tac" are two pieces of furniture, ideal for tea and coffee breaks. Taking a comfortable seat on "tic" and setting your hot cup on the attached table activates patterns hidden in the table's textile surface. These patterns – your "x" or "o" mark – are communicated to the textile surface in "tac". By intention or accident, you can discover and invite another into an aesthetic and subtle game of "tic-tac-toe" that lasts just as long as your coffee stays hot [7]. Reprinted from http://www.tii.se, with permission from Interactive Institute.

inks are not only used for decorative purposes [7], but also as indicators to warn for cold/high temperatures (Figure 6).

   Smart materials form part of a *smart system* that also has the capability to sense its environment or external stimulus and, if truly smart, to respond to that external stimulus and adapts its behaviour to it. The stimulus can be of various origins, which are illustrated in Figure 7.

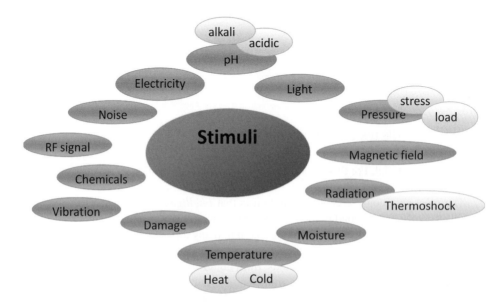

Figure 7. What triggers smart behaviour?

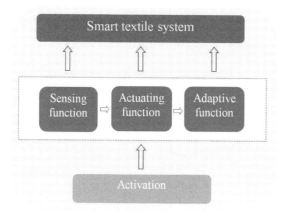

Figure 8. A smart textile system can sense its environment, act upon it and adapt its behaviour to it.

The definition of a smart system applies very well to a smart textile system. However, often the sensing function alone is taken as sufficient enough to constitute smartness in textiles. Due to the manner of behaviour, a smart textile system can be classified into three categories [8]:

• Textiles with a sensing function are referred to as *passive smart textiles.*
• Textiles with an actuating function are described as *active smart textiles* as they sense a stimulus from the environment and also act to it.
• Textiles with an adaptive function are called *very smart textiles* as they take a step further and have the gift to adapt their behaviour to the circumstances (Figure 8).

The textile may incorporate the following functions to build up a smart textile system:

• Sensing.
• Actuating.
• Powering/generating/storing.
• Communicating.
• Data processing.
• Interconnecting.

A smart garment, as depicted in Figure 9, does not necessarily contain all these six functions. The different functions may be in form of electronic devices or an inherent property of the material or the textile structure. Precondition, however, is the garments' capability to remain flexible, comfortable to wear, washable, durable and resistant to textile maintenance processes.

2.1.1. Sensing

A smart textile with a sensing function is able to "feel" changes in the environmental condition. A sensor is defined as a device providing information mostly in the form of an electrical signal. It senses the measured object or medium and emits a signal related to the variations of the measured quantity [9]. There are a lot of common sensors existing; however, for the application in smart clothing, piëzo-resistive [10–12] and pressure sensors

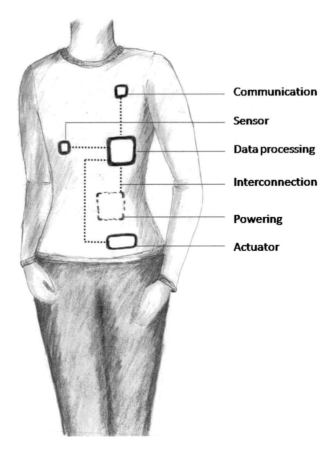

Communication

Sensor

Data processing

Interconnection

Powering

Actuator

Figure 9. The six functions of a smart textile system.

[13–15] are the most common types. The measured quantities are manifold ranging from position, velocity, temperature, humidity and force to pressure and flow (Figure 10). As most signals transmitted by the sensors are in electrical form, the most effective way to create a textile sensor is therefore by using electro-conductive materials.

2.1.2. Actuating function

The actuators' task in smart textiles is to react to the signal coming from the sensor or data-processing unit, respectively. The type of reaction may be in form of movement, noise or substance release.

Within the field of smart textiles, intensive research is led towards mechanical actuators based on shape memory materials or electro-active polymers [16–19].

Examples of chemical actuators are materials releasing specific substances such as fragrance and skin care products triggered among others by temperature, humidity or chemicals [20–22]. These substances can be bound chemically to a polymer fibre or stored in containers integrated into a fibre or put on a fabric in a coating layer. Thermal actuators in the form of electro-conductive material can be used as a heating or cooling element by exploiting the property of the material of having an electrical resistance.

Figure 10. A textile pressure sensor prepared during the Smart Textiles Salon 2009 in Ghent, Belgium. The pressure sensor is used as an on/off switch for the red LED. The sensor is connected to the light by conductive textiles serving as an interconnection.

2.1.3. Data processing

Until now a data processing unit made out of textiles does not exist. Electronic devices are still necessary to provide a computing ability.

2.1.4. Power supply and storage

Smart textiles require energy supply and energy storage capacity in order to function as a stand-alone unit. Other components, making up the smart textile system are activated through electrical power provided by power supply technologies. Flexible solar cells [23–25], micro fuel cells [26], flexible batteries [27–32] and the possibility of transforming body motion into electric power [33] may serve to provide electrical power.

Battery technology, for instance, has advanced over recent years because batteries are smaller, lighter, higher in energy and rechargeable. Some varieties are even mechanically flexible and water-resistant (washable). An example of a battery capable of providing electrical power for interactive electronic textiles was developed by NEC (Figure 11). It is a printing battery, which has a thickness of 0.3 mm and is rechargeable.

Flexible solar panels are successfully commercialized today, e.g. by Silicon Solar Inc. (Figure 12) and considerable research is also undertaken on the integration of flexible solar panels into textiles (Figure 13) [34].

2.1.5. Communication

Among the components of a smart textile (or garment) and between the garment and its wearer, communication is mandatory.

Wireless technologies, such as wireless network ports, eliminate the need to carry bulky processors and storage devices, simplifying the task of connecting electronics together. Especially in a medical environment it is necessary to communicate on a longer distance to provide immediate help in case of risky situations.

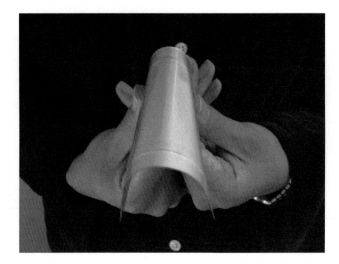

Figure 11. NEC's flexible battery that is designed for electronic paper [32]. Reprinted from http://www.nec.co.jp/press/ja/pr-room/bat-img/flexible_battery-01.jpg, with permission, courtesy of NEC Corporation. Unauthorized use not permitted.

The availability of electro-conductive textiles gives way to manufacture flexible antennas. These can easily be integrated into garments [35–37]. Among the variety of antenna topologies, Microstrip Patch Antennas, as shown in Figure 14, seem to be very suitable for this purpouse because of their compact geometry and planar structure. Such a textile antenna was developed by researchers of Ghent University. The antenna consists of a multilayer structure combining electro-conductive textiles for the radiating patch (antenna)

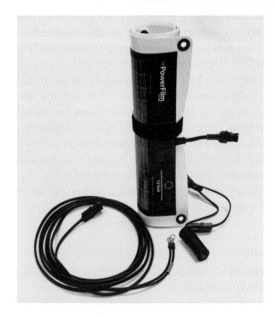

Figure 12. Commercialized rollable solar panels by Silicon Solar [25]. Reprinted from http://www.siliconsolar.com/visual-directory/12-v-consumer-ready-panels.html, with permission of Silicon Solar.

Figure 13. ScotteVest. The jacket has two small snap-on photovoltaic panels that fit onto its shoulders. These charcoal-gray solar panels convert the sun's rays into energy, which then feed a hidden battery pack about the size of a deck of cards. The batteries are wired to all the pockets, which can have almost any mobile devices plugged into them [34]. Reprinted from http://www.scottevest. com/v3_store/Fleece_Jacket.shtml, with permission of Scott Jordan.

and ground plane and conventional nonelectro-conductive textile material for the antenna substrate. Defining the electromagnetic properties of this substrate material was a crucial issue in the design process of the antenna, for which field simulators were used. Commercially available electro-conductive textiles and commonly applied textile production techniques can be applied to manufacture these antennas.

Communication between a smart garment and its wearer has been realised by France Telecom with an Optical Fibre Flexible Display (OFFD) [38]. Several prototypes like a backpack or a sweater come with integrated OFFDs. The flexible display is made up of plastic optical fibres forming a screen constituent of a number of pixels. The surface of the fabric is treated mechanically: micro-perforations are put into the fibre surface. When the light is emitted on the surface of the fabric, it scatters out and creates pixels. Figure 15 depicts another communication textile based on photonics – the Philips Lumalive fabric [39]. In this fabric, arrays of LED pixels were mounted on a flexible lightweight substrate, with each pixel containing close assembled red, green and blue (RGB) LEDs. Electro-conductive tracks on the plastic substrate connect these pixels into a matrix display configuration, allowing each pixel to be individually addressed and the light intensity of its three LEDs varied in order to produce any desired colour.

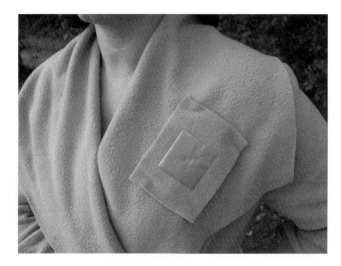

Figure 14. Microstrip patch antenna [36]. Reprinted from C. Hertleer, *Design of Planar Antennas Based on Textile Materials*, PhD Thesis (2008).

2.1.6. Interconnection

The electronic components that make up a smart textile system must be connected to each other in order to create versatile, interactive systems [40, 41]. Wires, cables and connectors are common physical materials used in the electronic world to connect electronics together (Figures 16 and 17).

Figure 15. Philips Lumalive [39]. Reprinted from http://www.lumalive.com/AboutUs/Press, with permission of Philips Lumalive.

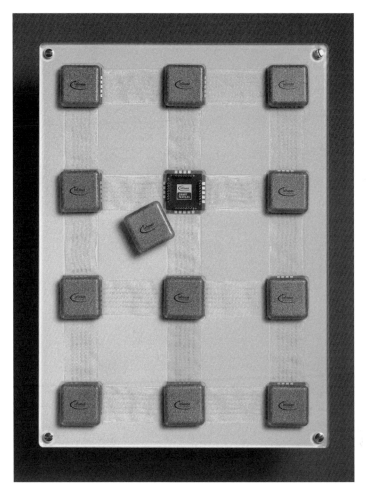

Figure 16. Scheme of an embedded electronic system in a textile fabric [40]. Reprinted from http://www.vorwerk-teppich.de/sc/vorwerk/template/bildemeldung_thinkCarpet_en.html, with permission of Vorwerk Teppich.

The interconnection or communication between different components making up a smart textile system is mainly realised by electro-conductive yarns woven into textiles [42, 43] to form a bus structure (Figures 18 and 19).

The techniques or contact mechanisms to join the different parts together are manifold. Embroidering, as shown in Figure 20, can be used for textile keypads and flexible modules [44]. Interconnections with electro-conductive adhesive and low melting solder material have also been shown [45], as depicted in Figure 21. These techniques can be accepted for clothing or other textiles as long as the module is reasonably small.

Sometimes it is desirable to interconnect electronics and textile substrate in a separate process [46]. For instance, a display may not be washable or could be used in different pieces of clothing. For that purpose, the usage of snaps is advisable.

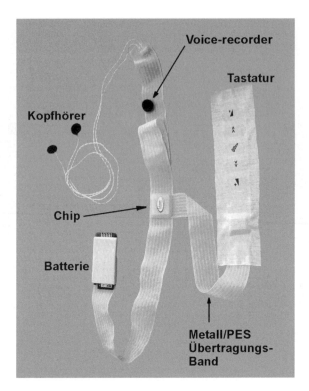

Figure 17. Prototype of an integrated textile system [41]. Reprinted from http://www.textile-wire.ch/downloads/neu_textile_wire_doc_de.pdf, with permission of Infineon/Interactive Wear.

3. Research and product landscape – where are we now?

3.1. Research

Currently, a lot of research around smart textiles is conducted worldwide. Most likely, it concentrates on the development of single components, e.g. sensors or actuators, making up a smart textile system, which in a later phase can be put together to shape a smart textile

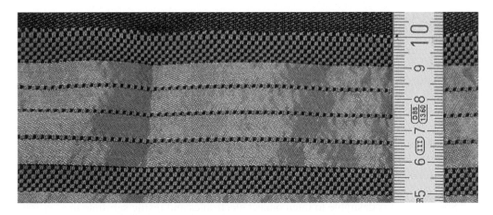

Figure 18. Textile bus structure developed by the TITV Greiz [42]. Reprinted from http://www.titv-greiz.de, with permission of TITV Greiz.

Figure 19. Electro-conductive ribbon sample that can be integrated into a jacket [43]. Reprinted from http://www.ohmatex.dk/images/lederbaand-030.jpg, with permission of Ohmatex.

system for protective clothing or medical textiles. For this roadmap we concentrated on gathering information on research initiatives that are undertaken in Europe, as they reflect the common research landscape worldwide. Hence, in this section a number of European projects around smart textiles are described.

Figure 20. Embroidered interconnections for textile keypads and flexible electronic modules developed by Fraunhofer IZM [42]. Reprinted from http://www.izm.fraunhofer.de/EN/About/strat_ allianzen/uni/HeterogenousTechnologyAllianceHTA.jsp, with permission of Fraunhofer IZM.

A. Schwarz et al.

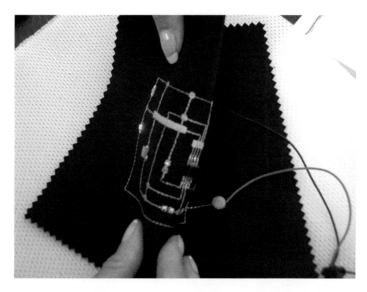

Figure 21. Embroidered circuit by TITV Greiz [45]. Reprinted from http://www.titv-greiz.de, with permission of TITV Greiz.

There are a number of European Commission (EC) co-financed projects, which contribute to the improvement of health monitoring based on textiles and garments according to a predefined scheme: it starts with "physiological parameter-oriented projects" (WEALTHY, MyHeart, MERMOTH, STELLA and OFSETH) relying on textile-embedded sensors, it adds biosensing (BIOTEX), and extends in the combination of the sensed signals (PROETEX) and beyond [47]. These seven projects form the cluster of Smart Fabrics, Interactive Textile (SFIT) and flexible wearable systems and are, next to a few others, described below.

3.1.1. WEALTHY

In the WEALTHY project smart materials in fibre and yarn form endowed with a wide range of electro-physical properties (conducting, piëzo-resistive etc.) were integrated and used as basic elements to implement wearable systems for collecting physiological data [48].

Integrated strain sensors based on piëzo-resisitive material as well as fabric electrodes were made of metal-based fibres. The piëzo-resisitive material consists of a Lycra® fabric coated with carbon-loaded rubber and a commercial electro-conductive yarn. The metal-based fibres for the fabric electrodes comprised two stainless steel fibres twisted around a viscose textile yarn. Above connections have been made out of textiles, using the circular knitting technique.

The system can monitor physiological variables, such as respiration, electrocardiogram, activity, pressure and temperature. A miniaturised short-range wireless system has been integrated into the garment, which transfers signals to the WEALTHY box/PCs, PDA and mobile phones. The garment interface is connected with the PPU, where local processing as well as communication with the network are performed.

3.1.2. WearIT@work

The WearIT@work – empowering the mobile worker by wearable computing – project is aimed to prove the applicability of computer systems integrated to clothes, the so-called wearables, in various industrial environments. Computer systems support their users or groups of users in an unobtrusive way, e.g. wearing them as a computer-belt. This allows them to perform their primary task without distracting their attention, enabling computer applications in novel fields [49].

3.1.3. MyHeart

The aim of MyHeart project was to gain knowledge on a citizen's actual health status by continuous monitoring of vital signs. The consortium consisted of 33 different partners from 11 countries. It integrated system solutions into functional clothes with integrated textile sensors. The combination of functional clothes and integrated electronics capable of processing them on-body can be defined as intelligent biomedical clothing. The processing consisted of making diagnoses, detecting trends and reacting to them. the MyHeart system was formed together with feedback devices, which were able to interact with the user as well as professional services.

This system was suitable for supporting citizens to fight major cardiovascular disease (CVD) risk factors and help to avoid heart attack and other acute events by providing personalised guidelines and giving feedback. It provides the necessary motivation to adopt the new life styles [50].

In order to sense different body parameters, scientists at the Wearable Computing Laboratory at the ETH in Zurich developed together with the company Sefar Inc. a System-on-Textiles, which is a woven fabric with thin insulated copper fibres.

In order to create an arbitrary conductor path within the textile, single copper wires must be connected at crossing points. This connection forms a textile via the fundamental building block for a connecting structure in fabrics. As there is a linear dependency of temperature of copper wires and their electrical resistance, the fabric is capable of measuring temperature. Due to the usage of the wires as warp and weft material, a grid like structure is formed, which enables it to locate the hot spot by measuring the resistances of warps and wefts.

For the purpose of sensing pressure, the researchers developed a pressure sensor mat made of a spacer fabric with embroidered electro-conductive patch arrays on both sides. With this system, the sitting posture can be detected. Each opposing patch pair in the array forms a plate capacitor whose capacity changes with compression force on the spacer fabric.

Finally, a prototype T-shirt with textile and rigid off-the-shelf sensors was developed. Within the scope of the European MyHeart project, researchers at the Wearable Computing Laboratory are working on an automatic dietary monitoring system. At the conference UbiComp in Japan in 2005, they presented their first results. They demonstrated that sounds from the user's mouth can be used to detect that he/she is eating and even that different kinds of food can be recognised by analysing the chewing sound. So far, the sounds are acquired with a microphone located inside the ear. However, in future they want to investigate other components of the dietary system, among them are a collar electrode for the detection of swallowing motions [51].

3.1.4. MERMOTH

The MERMOTH project contributes to medical remote monitoring of clothes. The objectives of the project are on the one hand to design a combined textile/hardware and software architecture for a family of wearable clothes, which provide continuous ambulatory monitoring of patients in academic research and the clinical trials of drugs and on the other hand to build prototype sets of garments, which address the two applications market with different compromises between power/distribution and consumption, user friendliness, relevance of the collected data and cost of the ownership [52].

3.1.5. Avalon

The project Avalon is aimed at the cross-sectorial development of hybrid textile structures integrating multifunctional shape memory alloys (SMAs) and the related processing techniques as well as design, simulation and organisational methodologies. The integration of such textile structures into novel high performance products in the fields of smart wearable systems and textile reinforcements for technical applications was realised [53].

3.1.6. Biotex

The overall goal of the Biotex project was to create a garment that monitors biochemical parameters of the wearer. For this purpose, the project aimed first at developing patches, adapted to different targeted body fluids, such as blood and sweat, and biological species, where the textile itself is the sensor. The technology was extended to the entire garment, as depicted in Figure 22. The project can be seen as an extension of the two former projects

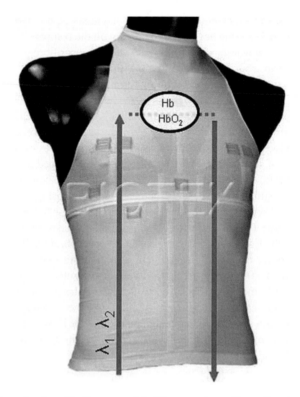

Figure 22.   Prototype developed in the scope of the Biotex project: shirt performing reflective oximetry using plastic optical fibres [54]. Reprinted from http://www.biotex-eu.com/html/results.html, with permission of CSEM.

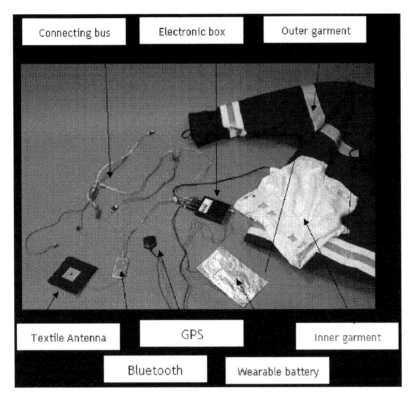

| Connecting bus | Electronic box | Outer garment |

| Textile Antenna | GPS | Inner garment |
| Bluetooth | Wearable battery |

Figure 23. Proetex prototypes of inner and outer firefighters garments [55]. Reprinted from http://www.csem.ch, with permission of CSEM Centre Suisse d'Electronique et de Microtechnique SA.

WEALTHY and MyHeart, in which physiological parameters, like ECG, temperature, movement and respiration were monitored [54].

3.1.7. ProeTEX

In the ProeTEX project a full system for firefighters and civil protection workers (Figure 23) plus a limited system for injured civilians were developed. Hence, the focus of the project is made on textile-based micro-nanotechnologies within a communicating framework. Textile- and fibre-based integrated smart wearables for emergency disaster intervention personnel with a goal of improving their safety, coordination and efficiency and additional systems for injured civilians aimed at optimising their survival management were developed. A wearable interface for monitoring the operator's health status and potential risks in the environment was developed. Thus, the operator obtains useful real-time information and/or alarms. Above that, data transmission between the operator and the central unit is created [55].

3.1.8. Stella

The Stella project aimed at developing integrated electronics in stretchable parts and products with stretchable conductors for use in healthcare, wellness and functional clothes

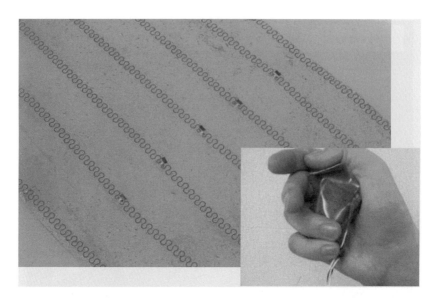

Figure 24. STELLA stretchable circuit board [56]. Reprinted from http://www.stella-project.de/, with permission of Thomas Löher, TU Berlin.

(Figure 24). For this purpose the consortium modified existing elastomers and non-woven substrates and developed conductor patterning and interconnection methods [56].

3.1.9. OFSETH

The aim of the OFSETH project was the integration of optical fibres-related technologies into functional textiles to extend the capabilities of wearable solutions for health monitoring. Two main technologies, one based on Fibre Bragg Grating sensors and the other based on Near Infrared Spectrometry, were used. These technologies can assess various parameters such as cardiac, respiratory rates (Figure 25) and oximetry that were investigated, and also pH or glucose concentration. In general, target applications were based on wearable static sensors to monitor cardiac and respiratory activities as well as oximetry, and wearable mobile sensors to demonstrate the wearability of Fibre Bragg Grating (FBG) and fibre near infra red spectrometry (NIRS) monitoring [57].

3.1.10. Lidwine

The Lidwine project focuses on the development of multifunctional medical textiles to prevent and treat decubitus wounds. Target applications include an antibacterial textile for wound care, integrated with medication depots including an active circulation support bandage. Therefore, active (controlled contractive cuff) and passive (textiles with integrated electrodes) systems are developed and applied. Further, drug capsules with an enzyme-based on/off switch for release are created.

3.1.11. INTELTEX

The overall objective of the INTELTEX project is to develop a new approach to obtain intelligent multi-reactive textiles integrating nano-filler-based electro-conductive polymer

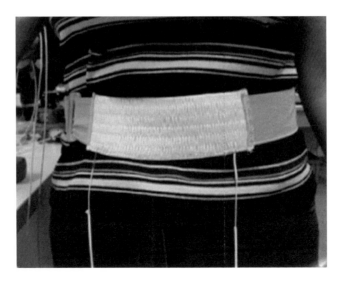

Figure 25. Example of a prototype of the respiratory motions sensor based on fabrics with embedded optical fibre [57]. Reprinted from http://www.ofseth.org/, with permission of MULTITEL
(www.multitel.be), project co-ordinators of the FP6 IST2004-027569 project.

composite (CPC) fibres. The consortium works on various scientific and technical activities,
ranging from CPC synthesis, the control of carbon nanotube dispersion in polymers to
fabrication of CPC fibres and textiles. In a subsequent step, the developed materials are
used in specific applications including protective clothing, anti-theft devices and textiles
for buildings and health [58].

3.2. Products

"Where are the products on the market?" is a frequently asked question when talking about
smart textiles. Despite around 10 years of research and development we only find a few
smart textile and clothing products on the market. However, we clearly need to distinguish
between component parts, such as electro-conductive yarns [59–61] or sensors, and more
sophisticated products, such as medical shirts measuring different body parameters, which
incorporate a whole system.

Electro-conductive yarns, for instance, have already been in the market for some years
and their development has followed a traditional path. As new electro-conductive materials
become available, fibre manufacturers find ways of converting them into yarns and fibres.
There are already several competitors in this market (Figures 26–28).

The more sophisticated component products, like textile pressure sensors, are made by
a small number of companies and the competition is weak.

The British company Eleksen, for instance, commercialises a soft and flexible textile-
based sensory fabric under the trade name ElekTex® Smart Fabric Interfaces [62]. It is a
combination of electro-conductive fibres and nylon. The principal applications are focussed
on XY positioning and pressure measurements. The point where pressure is being exercised
on the fabric can be localised by means of this XY positioning. This technology is already
applied to make "soft" telephones and a folding keyboard. For instance, the smart fabric
touchpad controller can be integrated into bags or backpacks for MP3 players to remotely
control music functions or into car seats to adjust the sitting position.

Figure 26. Stainless steel yarn from Bekintex [59]. Picture by author.

Products with partly integrated systems are expensive and mainly address the enter-tainment and healthcare markets. The most prominent example for an smart textile product can be found in the entertainment/communication market; it is a jacket with an integrated textile control panel for operating an MP3 player or a developed mobile phone. Among various manufacturers, Interactive Wear was one of the first companies that introduced this technology integrated into a jacket (Figure 29) [63].

These kinds of clothes received a lot of attention and publicity, but the economic success is still lacking.

When taking a look at the healthcare market, we find a few other smart textile products. The NuMetrex™ system is one example. The system concerns a sports bra with an integrated textile sensor to measure the heart rate of the wearer [64].

Figure 27. Silver-coated nylon yarn from Statex [60]. Picture by author.

Figure 28. Amberstrand® – metal coated Zylon yarns from Syscom Advanced Materials [61]. Picture by author.

Sensor technologies are not only integrated into garments but also into shoes. The company Adidas markets a sport shoe (Figure 30) that analyses the wearer's speed, weight and the terrain underfoot by an electronic unit implanted in the arch of the sole [65].

Further, heatable textiles bring a few more products around. One example of a heatable textile product is the MET 5 jacket by North Face (Figure 31). The jacket is constructed from a tightly woven, nylon fabric with stainless incorporated steel yarns that conduct heat; it is manufactured and licensed by the US company Malden Mills [66]. The fabric is powered by rechargeable lithium ion batteries, and is controlled from a pliable switch panel located on the upper left chest of the jacket and welded right into the fabric (batteries and AC recharger included).

Another commercialised product is the so-called WarmX [67] undershirt (Figure 32). The shirt is knitted with integrated silvered fibres in the kidney or neck zone. The silver-coated polyamide fibres warm up directly on the skin, power is supplied by a small battery positioned in a small waist pocket at the front of the shirt. The power controller is connected to the electro-conductive fabric through two snap fasteners. Thus, the power controller and the electro-conductive fabric form an electric circuit around the waist.

The above-described products give a brief overview on the products that can be purchased from the market today. It can be seen that the commercial importance of smart textiles has begun to be recognised. This perception may be based on the fact that smart textiles may be one key to competitive advantage over low price imports for the textile industry.

In order to get a quick overview of the development stage of the different smart textiles and their evolution, we defined five different levels of development, starting from research, going to prototyping, certification, commercial introduction and finally arriving at technical

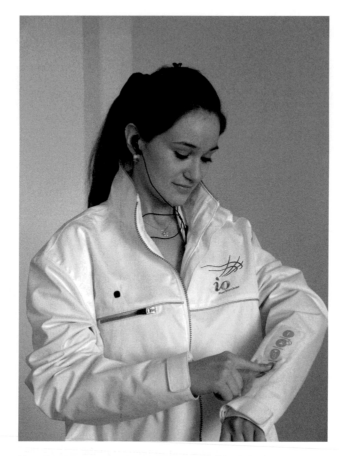

Figure 29. Multimedia jacket with integrated keypad to operate a mobile phone or listen to MP3 songs [63]. Reprinted from http://www.presseagentur.com/interactivewear/detail.php?pr_id=832&lang=en, with permission from Interactive Wear.

maturation and wider commercial use. Above that, the different levels of development help to learn about the marketability of special smart textile applications and the expected time-to-market, the stage of maturity need to be elaborated.

Figure 30. ADIDAS 1 shoe [65]. Reprinted from www.adidas.com/campaigns/adidas_1/content/index.asp?strCountry_adidascom=nl&strBrand_adidascom=performance, with permission of Adidas Benelux BV.

Figure 31. MET5 jacket by North Face using Polartec® heat: heating packs are inserted into a pouch in the jacket and the heating element is not directly sewn into the garment [66]. Reprinted from http://www.polartec.com/, with permission of Polartec.

3.3. Stages of maturity

3.3.1. Research

Applications in this phase have received interest for at least one or more researchers in the world. Some applications might be still in early development as they are tough to develop and need a lot of research to be fully understood. Many applications are currently in this phase as researchers are still struggling to improve the performance and stability.

3.3.2. Prototyping

After the hypothesis is validated, research typically but not necessarily moves from pure research labs to more applied research/commercial labs and companies. Applied research and development will eventually result in a proof of concept, a successful demonstration model. While the production issues might not have been solved yet, a successful proto-type/model has been validated. This is the stage in which various EC-funded projects are situated for the moment.

3.3.3. Certification

The development is starting to prove itself, the movement and transition towards real-life applications is moving forward and the demand through the industry is beginning to increase. An example to point out is the ECG sensing bra by Numetrex.

Figure 32. Electro-conductive fabric area of the warmX®-undershirt and the system components [67]. Reprinted from http://www.warmx.de/, with permission of WarmX.

3.3.4. Commercial introduction

After first demonstrator models and prototypes, initially, a usually prohibitively expensive, small number of products may be produced. All the same, if these prove successful, companies will seek to enhance production to gain market share. Generally at some point when feasibility has been proven production has to start. The product is now ready to be available for the end consumer, still not everywhere and at a rather high price. Jackets with integrated keypads, as described earlier, can be classified in this stage.

3.3.5. Technical maturation and wider commercial use

In this final development phase, production has reached significant numbers and research focuses on incrementally improving the products.

The textile industry is a small and medium enterprises (SME)-driven sector. Thus, a single SME has a certain position in the textile supply chain but does not cover the whole supply chain. The above-described stages of product development have different order of importance for companies depending on their position in the textile supply chain. Three main categories can be defined:

- Developer of smart textiles, e.g. consultancies.
- Producer of smart textiles, e.g. yarn manufacturer, finishing company, knitter and weavers.
- User of smart textiles, e.g. clothing companies, sportswear companies.

These types of companies have a special interest in specific smart textiles according to their research and development budget, resources and position in the supply chain (Figure 33).

Smart textile developers have special interest in materials which are at the development level of *research* or *prototyping*. As these companies have their own research and development laboratories, they are interested in further developing and finding solutions. They are technology-driven and used to find technological solutions.

Smart textile producers focus on the materials which are at *prototyping*, *certification* or *commercial introduction*. These companies are not interested in further development of smart textiles but rather have competencies in the production and manufacturing. The focus moves away from development technologies towards manufacturing technologies. Process knowledge and facilities are most important. On top of that, these SMEs have a network of recipients and the focus is moving towards customers.

Smart textile users may only be interested in smart textiles that are at the levels of *commercial introduction* or *technical maturation and wider commercial use*. They do not have the interest or the potential to develop or produce smart textiles. These SMEs are customer-focussed and provide end consumers with solutions to their problems. They do this with the support of smart textiles that can be purchased, e.g. electro-conductive yarns, and integrated into their value-adding processes.

Certainly, most efforts are currently concentrated on the development technologies, meaning research, prototyping and certification. The first prototyping and the subsequent certification, in most cases, did not receive the needed maturity yet to proceed to the next step, i.e. the *commercial introduction*.

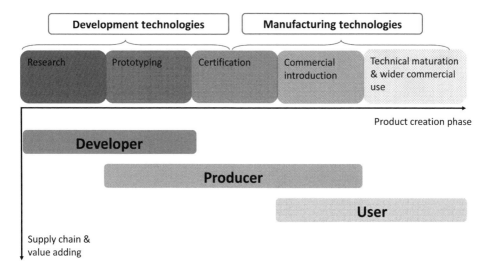

Figure 33. Relevant product development stages for different SME types.

However, it gets apparent that the key opening for smart textiles is in *product innovation*, rather in *process innovation*. New materials will be developed, or improve the properties of existing materials, e.g. antistatic yarns in carpets, instead of trimming down the production costs or improve quality.

4. Smart textile market potential – where do we want to be?

Despite considerable publicity and many years of research and development, very few smart textile products have been successfully commercialised at any volume as described above.

In most of the few smart textile products that are commercially available at this moment, the smartness of the textile component is still rather low. By integrating technologies the smart textile product is being made up by microelectronic components, while the textile itself serves as a backing, assuming its traditional functionalities for any given application [68]. One example to enumerate at this point again is the integration of the iPod, headphones and Bluetooth-compatible microprocessors in various textile-based products.

However, taking a look at the current research activities in the field of smart textiles, which have rapidly being intensified over the past years, it gives reason to expect that the situation is improving. Novel textile technologies, such as textile antennas, textile strain gauges, polymeric fibres able to deliver controlled doses of therapeutic agents and fibres that capture and transfer solar energy, are being developed.

Besides the growing research activities, the forecasting studies presented in the following paragraphs give reason to expect a growing smart textile market.

The analysis and synthesis of the smart textile market potential is based, on the one hand, on the opinions of the limited number of international experts gathered through a Delphi method, and on the other hand on the market forecast data presented in Section 4.2, which are extracted from available market-forecasting studies on smart textiles.

4.1. Forecast through a Delphi study

4.1.1. Methodology

The Delphi study carried out in the scope of this roadmap provided a strategic analysis allowing the identification of prioritised needs and potential gaps for the transition of textile industry to a knowledge-based industry.

This study has been inspired by and adapted from the conventional Delphi method, involving the use of a series of questionnaires designed by a monitor group and then sent to 210 experts. The experts were split up into the following two groups:

- Scientific experts, knowing the technologies involved in the production of smart textiles.
- Industrial experts, knowing the textile industry and its main client sectors.

The scientific experts provided information in terms of scientific relevance, feasibility and position of the EU research, while the industrial experts provided their opinion in terms of business relevance, acceptance criteria and socioeconomic expected impacts. The first group had the mission to stress what is scientifically and technically feasible, while the industry group had to provide a more global view of the innovation process and awareness on essential technological and non-technological aspects of the innovation.

In total, two question rounds were organised. The first round of questioning provided a basis to view the potential impact of smart textiles development. In total, 43 exploitable questionnaires have been collected during this round.

On the basis of the received answers the experts were asked to answer additional questionnaires in the second round in line with their fields of expertise. In this round, the experts gave their views on the market development and research perspectives in the field of smart textiles. In total, 87 exploitable questionnaires have been collected for the second round. The aim was to create a consensus and a collective view of the future of smart textiles and to achieve the involvement of key actors in Europe in the designing of scenarios for the future of the textile and clothing sector, which have been reported above in last section.

4.1.2. Results

The results presented below are taken from the second round of questionnaires being sent to identify scientific and industrial experts.

To forecast the market of smart textiles, industrial experts were asked to estimate the application fields of smart textiles, share of different materials leading to smart textiles in the three textile branches, namely clothing, interior and technical textiles, as illustrated in Figure 34.

The *clothing sector* represents the most prosperous application area among the three textile industries for smart textiles. Here, all materials except for advanced materials such as shape memory alloys are expected to be used extensively.

Encapsulating materials, such as phase change materials (PCMs) incorporated in micro-capsules or odour-catching components grafted to textile materials, and advanced polymers including colour change and shape memory polymers will be applied the most followed by electro-conductive yarns, sensors and actuators and energy-generating and storing textile

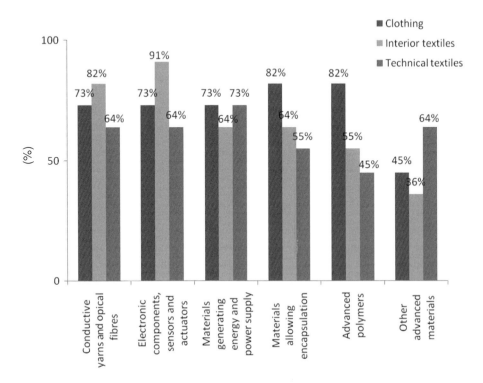

Figure 34. Forecasted application fields according to Delphi study.

Figure 35. A jacket with integrated Outlast® technology [69]. Reprinted from http://www.outlast. com/index.php?id=1&L=0, with permission of Outlast.

materials to make up an intelligent garment. A link can be made to the actual market situation. PCM materials are increasingly used in jackets (Figure 35) and shoes to keep the wearer warm. Numerous brands of commercialised garments with integrated microcapsules are available [69].

It is a bit surprising that the experts believe that advanced polymers will find more applications in clothing than electro-conductive materials and electronic components. Taking a look at the clothing market today, it can be seen that various clothes with integrated control panels and switches to operate, for instance an MP3 player, are marketed. On the contrary, clothing products that dress up by advanced polymers rarely exist so far.

It stands to reason that the last group of materials including ceramics and shape memory alloys are likely not to be used in clothes, probably due to their poor mechanical properties and the high degree of stiffness they possess.

For *interior textiles*, electro-conductive materials and electronic components will be predominantly used. This may indicate that interior textile will be mainly equipped with sensory and lighting functions. Above that, materials generating and storing energy and materials for encapsulating will be in demand. Taking a look at the product and research palette being offered today, it can be seen that those numbers are in good accord. Research, for instance, is going on in developing curtains that collect sun light and release it in the form of heat or cold [70]. Above that, prototypes of carpets with integrated sensors and lightening fibres [40] exist next to pillows with integrated control panels serving as remote control for the TV [71] and textile wall panels that change their patterns [72] (Figure 36).

Finally, advanced materials such as shape memory alloys and ceramics have the slightest prospects in this domain.

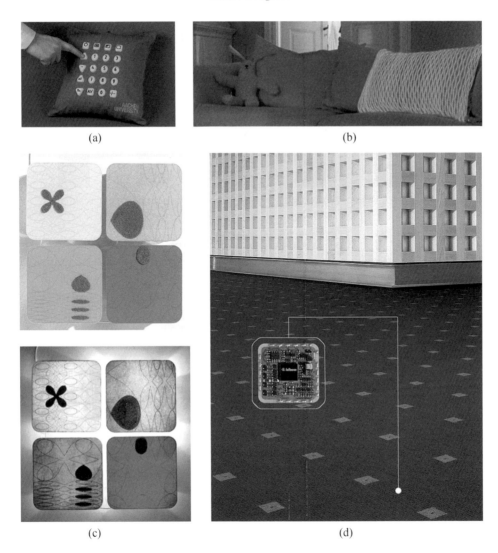

(a) (b)

(c) (d)

Figure 36. Intelligent textile products: (a) Pillow with remote control for the television [71]. Reprinted from http://www.ita.rwth-aachen.de/ita/3-f-und-d/3-01-08-smart-textiles.html, with permission of ITA. (b) Interactive pillow with electroluminescent wires [7]. Reprinted from http://www.tii.se, with permission of Interactive Institute. (c) Fuzzy Light Wall [72]. Reprinted from http://www.ifmachines.com/, with permission of Maggie Orth. (d) The thinking carpet developed by Vorwerk and Infineon has a controlling alarm, climate control, regulatory or guidance systems technology [40]. Reprinted from http://www.vorwerk-teppich.de/sc/vorwerk/template/bildemeldung_thinkCarpet_en.html, with permission of Vorwerk Teppich.

Seemingly, smart textiles will be least utilized in *technical textile* applications when compared to clothing and interior textiles. Here, the highest expectations can be documented for energy supply and generating materials. It also gets apparent that other advanced materials like shape memory alloys have the brightest outlook in technical textiles. This result may be attributed to the fact that in some technical textile applications, performance is demanded more than textile-related properties, such as handle and

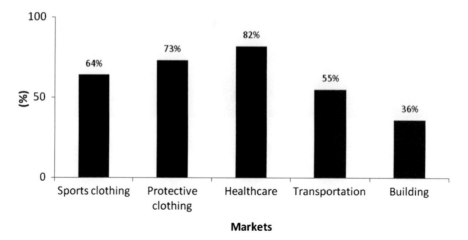

Figure 37. Forecasted market for selected applications according the Delphi study (remark: only electronic components are taken into consideration).

drape, and consequently stiffer and more brittle materials have a better future in this domain.

In general, it can be observed that for all the three textile sectors, electro-conductive yarns as well as electronic components have the brightest applications in the textile sector. Figure 37 indicates that the healthcare market will be the market making the greatest use of smart textiles for electronic components, followed by the protective clothing market. These materials are also supposed to be the first materials that will make up 10% of sales in the three textile sectors within the next 10 years, followed by materials with incorporated microcapsules (see Figure 38).

It gets apparent that materials generating and storing power will need more than 10 years to reach a representative market share. Additionally, advanced polymers will need more than 10 years from now to achieve respectable sales. Polymers with some kind of

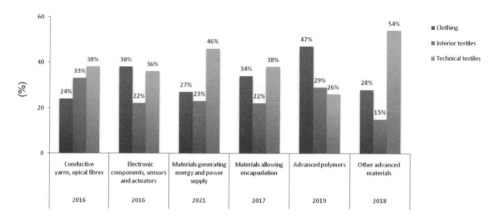

Figure 38. Forecasted split of sales (%) of smart textiles between the three selected sectors – clothing, interior textiles and technical textiles – according to Delphi study.

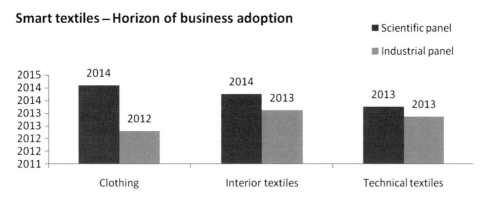

Figure 39. Horizon of business adaption according to Delphi study.

active function (e.g. electrochromic, piezoelectric, shape memory polymers) have potential applications in a wide range of end users but the current markets are small.

A major potential opportunity exists in clothing, as they will have a huge contribution to the advancement of clothes. Material requirements are often likely to be device-driven. In some cases (e.g. electronic polymers) requirements for smart textiles are unlikely to be the major commercial driver for the development of such products but merely a spin-off from their main applications.

In general, it gets obvious when looking at the numbers that the best sales will be probably achieved with technical textiles.

The horizon of business adoption was also evaluated in the study (Figure 39). It can be clearly observed that industrial people compared to the opinion of scientific people believe that smart textiles will be available much sooner in the market. The optimism of the industrial people may be related to the less precise view they have on current research activities or their wish to have smart textiles quickly implemented in the market.

Despite the huge expectation from smart textiles, the development is still at its early stage and there are lot of challenges to be met.

In most cases, end users are widely spread and potential demand is highly fragmented.

4.1.3. Key processes and technologies

Taking a look at the forecasted key processes and technologies that will be used to produce smart textiles (Figures 40 and 41), it gets apparent that finishing processes are most promising. A reason for this point of view might be that these processes can be adapted for the production of smart textiles. It is surprising that the scientific interest in spinning is rather low. When taking a look to present research activities, considerable research is undertaken, for instance, in spinning of polymer fibres loaded with electro-conductive particles [73,74].

4.2. Other forecasting studies

Several organisations forecasted the market development of smart textiles, which, according to these reports, is supposed to grow rapidly. Different forecasting studies propose very distinct figures, which casts doubts on the reliability of these studies. Certainly, it is difficult to forecast the market potential for smart textiles as there is still a lack of fundamental

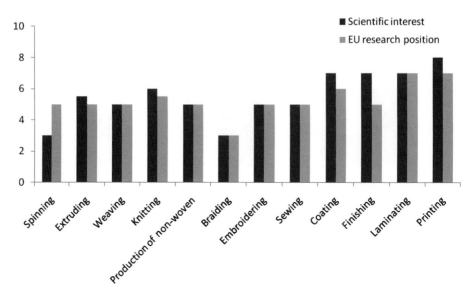

Figure 40. Forecasts on processes involved in smart textile production according to Delphi study.

information required to evaluate the situation, especially concerning the demand on smart textiles (Table 1).

4.2.1. Forecasted application areas

Next to the market potential, forecasting studies also predicted their application fields. A list with the application fields according to the different market studies is given below.

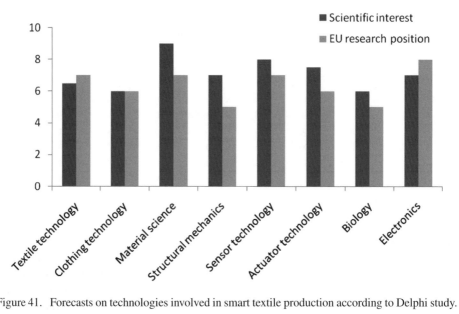

Figure 41. Forecasts on technologies involved in smart textile production according to Delphi study.

Table 1. Overview of different market forecasts (in million $).

Year	Global market		US market	
	VDC study (2007)	VDC study (2004)	BCC study	VDC study (2007)
2004		248	64.4	
2005		304		
2006	369.1	308	70.9	143.7
2007	440.9	486	78.6	175.2
2008	553.4	642		225.1
2009	747.3		299	318.3
2010	1129.6			532.5
Growth rate (%)	30	27		38.7

BCC study [75]:

- Consumer entertainment.
- Safety and protective clothing.
- Biomedical applications, including health status monitoring.

Frost and Sullivan [76]:

- Healthcare applications (remotely monitoring parameters).
- Security (detecting danger and calling for help).
- Display of helpful data (communication through the Internet or communication between people).

VDC study [77]:

- Clothing with built-in electronic storage.
- Clothing with communication capabilities.
- Fabrics with provide electromagnetic shielding.

According to these studies, the number of applications for smart textiles, which have made a commercial impact, is disappointingly low – apart from electrically heated seat kits, which have been a major success. End uses that have proven to be successful tend to be restricted to smaller markets such as luxury sportswear and novelty outerwear. The studies forecasted that an increasing demand on smart textiles is expected for the healthcare and the communication/entertainment markets. An indication for this prediction is that there are health- and fitness-monitoring products in the market, as well as clothes with integrated control panels for MP3 players and phones.

4.3. Technological evolution and scenarios

Smart textiles will be needed to combine a wide range of technologies. Since it is still at an early stage, its evolution is hardly to characterise or predict as it can be seen by the different forecast studies. Assuming that the benefits of smart textiles which they offer to society and industry will exceed the risks and possible negative effects, a great number of materials and technologies used to create smart textiles as well as the products themselves will be well defined and characterised in a standardised manner. In the following paragraphs, potential application scenarios are presented giving ideas on the opportunities and influences of smart textiles in our everyday lives in the years 2020 and 2030.

4.3.1. Scenario 1 – elderly people and interior textiles in 2020

Anne is a 98-year-old lady. She lives alone in the suburbs of London. Although she is alone, she feels safe in her comfortable apartment where the lamps light up naturally when she goes through rooms or corridors.

She also appreciates her new armchair with its textile keyboard that gives her the functionality of a remote control for her TV, and the dial-and pickup buttons of a telephone, the speakers being positioned inside the headrest. Of course, every important telephone number is stored!

Growing older, due to her low blood-pressure, Anne feels dizzy every now and then. Yesterday, returning home from shopping, she had to take a short rest and sat on the floor. A few moments later, she received a call to ask her if she was all right! Luckily the carpet had detected her position on the floor.

The cardiac and respiratory monitoring showed that she had nothing severe. The caretaker of her building helped her back on her feet.

Luckily, she was not obliged to live in a retirement home far from her district where she had lived for so many years. Later this evening her grandchildren visited along with her great grandchildren.

4.3.2. Scenario 2 – healthcare and anti-headache cap in 2020

In my youth I often noticed mother taking pills for pain relief. I swore that as an adult I would never use pills. Unfortunately, I inherited my mother's sensitivity to headaches. So I ended up being the same pill-taking person I never wanted to be. Last week my doctor invited me to try something brand-new: a cap that could reduce my headaches. I was very sceptical, but willing to give it a try.

Two days later I felt the headache approaching, but since I was still at work I had to wait until the evening to try my new cap. At my arrival the pain was so strong that I decided to try my cap immediately. It looked nice and felt comfortable. The first minutes I had a very pleasant sensation in my neck and forehead. The feeling spread over my entire head and believe it or not, it stopped the pain! I kept it on for the night, just to be sure the headache would not return.

I slept excellently and felt great next morning. Moreover, I'm closer now to the ideal of my youth: I finally live without taking pills.

4.3.3. Scenario 3 – protective workwear in 2020

MoreProtect is proud to present the new collection of warning workwear the LifeWear.

This incredible uniform is able to save the life and to avoid injures of your employees through an early detection–warning system.

Dangers that are detected are as follows:

- Harmful noises.
- Electromagnetic fields.
- Radioactivity.
- High voltages.

- Harmful gas.
- Overheating.
- Aggressive liquids and chemicals.
- Bacteria.
- Viruses.

Following are the advantages of the LifeWear:

- Elegant and smart uniform customised to your brand.
- Cost competitive solution.
- Standard protection against liquid, heat, shock, perforation etc.
- Very light and comfortable garment, thanks to integrated conditioned temperature and nanotechnologies.
- Auto-fit.
- Easy care.
- Durability.
- Autonomous (hydrogen capsule for a week).
- Light and sound alarms.
- Lighting messages, night lighting warning.
- Illness diagnostic.
- Drug delivery (five different life-saving drugs corresponding to the diagnosis).
- Radio and mobile phone communication with environment.
- Integrated Global Positioning System (GPS) (data and video/audio link).

4.3.4. *Scenario 4 – healthcare and surgery in 2030*

Patient: Michael, 18 years old.

Diagnosis: Excision of his skin and muscles of his left leg due to a car accident. Due to necrosis, soft tissue surgery was indicated.

Treatment: Removal of necrosis tissue; implantation of multiplayer implants as a substitute of muscles and skin.

Progress: Until 13 hours after implantation, a weak inflammation appeared at the place of implantation. No negative reaction was observed during healing process. Implant was accepted by the patient, no immunological reaction was observed. On the third day, the motoric rehabilitation was introduced. During convalescence period, patient had contact with staphylococcus aureus bacteria. Infection was stopped by implants without additional antibiotic treatment.

The patient left hospital feeling well.

Doctor's orders: Commence rehabilitation, control within three weeks.

Result: Percentage of health loss: 20%.

Discussion between Michael and his friend after treatment.

Friend: Hello Michael! How nice to see you well!

Michael: Hello! I am not well since they will send me less money as I expected. My health loss is only 20%. It is due to these new types of implants.

Friend: Are you crazy!? Do you prefer more money over a good health?

Michael: Just joking! I am really glad that my body almost gained full functionality. I was afraid of not being able to go skiing or drive my bike. Now I assume that will be possible quite soon.

Friend: That's great news!

Michael's mother diary.
10.03.2030
My dear diary,
Today Michael had an accident. His leg is almost destroyed. Parts of his skin and muscles are cut. I don't know what to do!!
. . .
12.03.2030
My dear diary,
I was so afraid about my son. After this accident I was sure, that he would have some problems to move. I thought that he wouldn't be able to walk or run! His surgeon found a new type of textile implants which allow him to move. I don't understand everything, but part of this textile implants will disappear when the original tissue grows. But part of it will stay inside, stabilizing the tissues.
. . .
20.03.2030
My dear diary,
Yesterday it appeared that in hospital there is risk of bacteria infection. Michael had a contact with another patient infected by bacteria. Yet the Doctor said that there is no reason to give him additional medicine, since it is included in the implant. If the infection will appear, the drugs will be released form implant to the place of infection.
. . .
28.03.2030
My dear diary,
My son is at home. He can walk, and his doctor said, that he will feel better every day.

4.3.5. Scenario 5 – fashion in 2030

A quote taken from the diary of a young Brazilian girl, 16 years old.
Johana's Web diary.
San-Paolo, June 12, 2031.
This morning I checked my favourite actress site, Intacha, she is a marvellous Spanish actress. I am fan of all her movies and I don't miss any news about her life. If I study well Dad promised to pay plastic surgery to transform my body to make it look like hers.
Two days ago I have seen her new movie "The heart of the oak". She was wearing a marvellous green dress. I found it, it is a CreaDress production available on [the] Internet, I have downloaded it in my MultiDress. After I tried it, I bought it for 53$ for a wearing-period of 72 hours. Of course there is automatic size adaptation.
I eagerly dressed up with my MultiDress and pressed the luminous start button on my hip.
Then slowly the dress took shape and the colour of the dress became green and blue and I was like Intacha in the movie. The MultiDress was moving slowly like when she was saying "good bye" to Lucas on the quay in the movie.
I have to show it to my friends Rosa and Juliet, they will be jealous.
The MultiDress is a garment able to change shape and colour by electronic signal (computer data downloaded). The most ambitious aspect should be to keep a good drape of the fabric, flexibility and lightness.

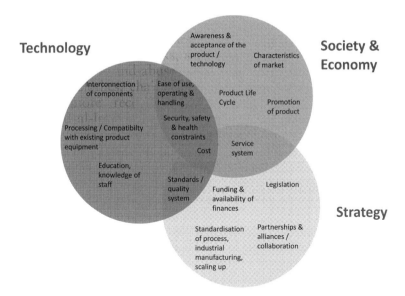

Figure 42.   Barriers for the progress in intelligent textiles are of technological, strategic and societal and economical origin.

As these scenarios describe future visions of our lives, we still have a long way to go since there are several technical and non-technical challenges to be solved. These will be described in the following chapter.

5. Barriers and drivers to progress and possible solutions – what is stopping and what encourages us getting there?

Smart textiles, being a relatively young domain, has to experience several gaps and barriers to overcome before becoming universally useable and acceptable. These barriers can be of technological, strategic or societal and economic nature (Figure 42). Technological focus is based around the issues of processing and fabrication as well as the compatibility with existing equipment, the interconnection of components into the system and the education and knowledge of staff. These issues are the major technological barriers to the exploitation of smart textiles.

It is also the practical implementation of smart textile technologies such as the awareness and acceptance of smart textiles, the ease of use, the service system, the promotion and life cycle of the product as well as the subsequent added value that contribute to an economic and societal advantage.

Further, the barriers that govern the development and commercialisation of smart textiles are also of strategic origin. If there is a lack of partnerships between intra- and intersectorial businesses and research approaches, then the availability of funding and finances is not guaranteed. Hence, the development of smart textiles will stagnate. Industrial growth is also inhibited by the lack of industry standards, which is hampering communication and technological progress.

Costs are a barrier that accounts for all the three dimensions of barriers and are thus discussed in the following subchapter. Subsequently, the technological, societal and economical as well as strategic barriers are described.

5.1. Costs

Costs are a major barrier in terms of technology, strategy and economy and society.

Let us first take a look to the *societal and economic barrier* cluster. Here, the market development of smart textiles is to some extent inhibited by the high cost of smart textiles. In general, smart textile products are offered at very high prices compared to their non-intelligent parent items, e.g. shirts and bras. Some of these products have required considerable development and the need to recover these costs is reflected in their selling price. An example is a health-monitoring garment, where it is clear that the costs are not the costs of the components but of the development of the signal-processing algorithms developed to filter out the noise. It is also likely that for many products the price is justified because the products aim at early adopters who will pay high prices for new gadgets. To go one step further and make the product attractive for the majority of consumers, it is very important that the cost-performance relationship is accepted by the market, because the commercialisation will only increase when the cost efficiency has been improved.

However, this is a delicate topic as there is no strong market pull for the moment. Without having a firm evidence of a market pull, companies are not willing to proceed to the next step, the further commercialisation, and finally that of higher cost efficiency. The question arises whether the products offered to the market for the moment will go beyond the early adopter acceptance or will just vanish from the market.

The high prices for end products at the moment are a consequence of high costs research and marketing, which are *strategic barriers*. A shortage of research funding and delays in funding counteract with high research costs. These factors lead to stagnation in research followed by inhibition of industry growth.

Additionally, the size of the company is also of great importance when considering cost as a barrier. In general, big companies are more focused on costs than SMEs because they are planning to achieve a cost leadership in markets with large volumes. For that reason, they are only pushing a product into the market when the technology is able to provide large-scale affordable solutions. SMEs are not so much focussed on costs because their special interest is to find niche markets for their products in order to survive and stay competitively in the market.

Large-scale affordable solutions are the keywords when thinking about cost as a *technological barrier*. The technologies used for the moment to produce smart textile products are mainly the adopted technologies used in the electronic industry combined with some textile technologies. As most of the technologies need to be adapted for smart textiles, they exist only on a laboratory scale, which enhances the cost of production. Some of these technologies are even difficult to be upgraded so that the market value needs to be questioned.

5.2. Technological barriers

5.2.1. Education, knowledge and training of staff

Its multidisciplinary nature makes smart textiles a challenging field for education and training of students and recruitment of the needed technical workforce.

Regarding higher education, students of electronics, chemistry, physics, material science and mechanical engineering seem destined to drive the smart textile development.

Education in this field is very important and necessary and takes place at most progressive universities' teaching institutions. Some pioneering initiatives can be pointed out, for example at Tampere University of Technology in Finland [78], at the Swedish School of

Textiles in Borås [79], at Ghent University in Belgium [80] and in the scope of the E-Team (European Masters in Advanced Textile Engineering) [81]. They offer specific courses on smart textiles.

So far the greatest effort has been done at Herriot Watt University in Scotland [82]. 2008, the university offered an entire master's programme in smart textiles and clothing systems with a focus towards the technological developments in smart textiles related to developing new products in applications such as personal health monitoring, personal protective equipment, construction, transport and geotextiles.

5.2.2. Standards and quality system

Quality improvement is a classical success factor in the domain of new materials. The improvement of material properties and the product quality gives direct competitive advantages to companies. Smart textiles concern niche markets with high-tech products, in which quality and material properties are essential to be successful.

For the moment, the growth is being inhibited by a lack of industry standards, which is hampering communication and technological progress. There is a pressing need for a common set of standards defining methodologies, specifications and best practices for the various smart textile applications and products.

There are standards for textiles and also for electronic components and devices available, but none that combine electronics and textiles together. The great need in a unified set of standards is imperative for the future success of smart textiles. Such standards would have to include polymers, electro-conductive fibres and application of cross-sectorial techniques. However, for the development of standards, the cross-sectorial communication, for instance between textile and electronic companies, must be improved.

Further, there is a body needed to oversee common standards and practices and to accept overall responsibility for these.

5.2.3. Interconnection of components

Taking a look to the existing prototypes and products in the field of smart textiles, it can be seen that the interconnection between different components of the smart textile system represents a major problem. This may be accounted to the fact that some components are already integrated in the textile form, such as sensors made of electro-conductive yarns, whereas some others, such as batteries, are still not available in textile-compatible form. In most cases, the interconnection between different components is still made by using electronic cabling, which is connected to different components by soldering, welding, embroidering or sewing.

Also in the area of makeup, problems exist in view of the critical need to position electronic panels correctly and not to damage them in the cutting process. Therefore, thorough layout planning is of crucial importance.

5.2.4. Others

Last but not the least, there is a lack of understanding the correlation between material, e.g. electro-conductive textile materials, and structural parameters. Simulation work would be beneficial to model, for instance, how electrical current is distributed in an electro-conductive textile or how a drug is released from microcapsules grafted to textiles.

5.3. Societal and economic barriers

5.3.1. Ethics

Ethics are an important feature to consider in our international and multicultural society. The technological progress is fascinating. Technology that can be integrated into our clothes will enhance the quality of our lives. It can be used to monitor and respond our body state and behaviour without supervision or interaction. Decisions that can influence and change our lives will be made based on these technologies. These decisions may not always have very positive purposes, but it can also be misunderstood or misused, creating some ethical issues that need to be pointed out.

5.3.2. Security, safety and health constraints

Nowadays life has become increasingly digitised and connected. More personal information can be digitally gathered and stored, and possibly other sources, services, institutions or persons can access our data. With smart textiles, especially electronic textiles, the monitoring capabilities can be massively extended.

However, this in turn raises the number of social and legal issues in relation to identity, privacy and security.

Addressing these issues will be the core challenge for the future of smart textiles. To create confidence in smart textiles, an acceptable level of security and privacy must be established. Privacy is at stake here and these information exchanges need to be secured and managed.

Possible health risks and constraints arise from the fact that electronics are integrated in textiles that are worn on the body or kept close to it: The consumer is exposed to radiation and hazardous influences due to current flow.

The Virtual Data Center (VDC) performed a survey among potential users of wearable computers regarding their concern for safety (Table 2).

In total 216 people were asked questions for the study. It indicates that people are concerned about possible dangers that may arise from wearable computers [77].

5.3.3. Promotion of product

There has been little discussion so far concerning potential new distribution channels for smart textile products which could be developed from conventional sales channels. Clothing retailers themselves may be sceptic towards smart textile products. This opens up issues such as guarantees and product liability in the face of extreme caution, in addition to insecurity regarding the demand of smart textile products.

5.3.4. Awareness and acceptance of the product/technology

It is known that innovations and changes of all types can result in scepticism, insecurity, disapproval and even confused anxieties, as well as curiosity. Technological developments

Table 2.   Safety concern of people regarding wearable computers according to VDC [76].

Very worried	Neutral	Little concern	No concern at all
27%	19%	12%	10%

are achieved quicker nowadays, making it more difficult to estimate their consequences. These often cannot be categorised in our existing day-to-day environment in the usual way. If, on the other hand, people are more objectively aware of the positive new opportunities offered by technological innovations and have a better understanding of the developments behind this, the necessary acceptance for technological fields relating to smart textile products could be created.

Players from the retail sector have not yet been sufficiently involved in the developing networks relating to the emerging technology. The presentation and explanation of new types of intelligent products at the point of sale can be regarded as a challenge.

On the other hand, textile companies need to be kept aware of the market requirements and technology developments.

The question arises, whether smart textiles are really needed even if sales people and consumers are not aware of them. Can a jacket with an integrated MP3 player really compete with the separate device in the long term, which provides the same solution?

The major drawback of smart textiles is that presently their reliability is insufficient. If a smart textile fails to react or function in normal manner, then it will be very difficult to get accepted by the public. It is more crucial to achieve a high degree of reliability and fault tolerance.

5.3.5. *Ease of use, operating and handling*

The necessity for smart textiles to be user friendly and easy to handle is relevant for both the technological as well as the societal factors. This signifies that there is a need to develop user-friendly products and services that are also perceived by the consumers as easy to handle. Related factors include the following:

- Reliability and predictability.
- Usability of technical devices by "non-technical" people.
- Configuration, personalisation and control by user.
- Friendly human–machine interface.
- Very simple communication protocols.

Factors correlated with the problem of universal access are as follows:

- Needs and capabilities of young people versus those of seniors.
- First world versus third world.
- Different needs in Central and Eastern Europe.

5.3.6. *Characteristics of the market*

Here, concerns over price and affordability as well as the needs for having adequate speed of market take-off (early adopters) and subsequently critical masses of users are encompassed. Certainly, it is still too early to talk about an existing demand for smart textiles. There is still no discernible market pull. This is particularly challenging for companies as the imperative needs for personalisation and customisation are also expressed ("mass-customisation").

The fundamental question that we need to put here is: where is the market for smart textiles to be found? Is it the textile and fashion market, the healthcare market, the electronics market, the sportswear market, or maybe all of them? This is a major question as so far there is no real market for smart textiles and it still needs to be evolved. The only branch where there may already be some smart textiles marketed is the automotive market. Here,

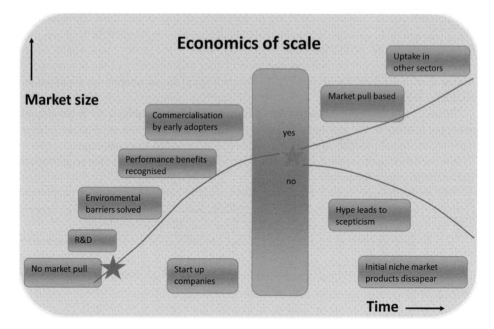

Figure 43. Economy of scale for a new product [83]. Reprinted from http://www.iom3.org/content/smart-materials-systems-foresight, with permission of National Archives Licence number P2010000269.

we can already find heatable car seats. Heatable car seats are still a luxury component, but in future may become a standard item for every car.

5.3.7. *Product life cycle*

The supply sector for smart textiles is highly diverse, ranging from electronics, materials and chemicals to healthcare, and different commercial pressures and drivers apply across the sector. Two possible scenarios, one positive and other negative, for market development in the supply sector are represented in Figure 43. Both assume an early growth stage, catalysed by research and development, and early successes where environmental barriers are overcome, performance benefits are recognised by innovative, early adopters, and start-up companies are established. However, the scenarios differ in one crucial element – in the extent to which economies of scale are achieved. Very often, early promises of new material development remain unfulfilled because costs cannot be driven down sufficiently to achieve volume growth, and cynicism can take hold, particularly if unrealistic predictions of the material technology were made in the first place. Within the EU, research funding is the normal mechanism in the early stages of such a scenario when there is little market pull, but great technical pull by the red star vision [83].

However, this schematic suggests that there is another critical point where government funding should be considered; and economies of scale are proved. This could ultimately be of greater strategic benefit in improving the success rate of start-up ventures in smart textile developments.

5.4. Strategic barriers

5.4.1. Funding and availability of finances

Financial investment for the development of smart textiles is high and involves high risks. A return of investment can only be achieved with high market volumes, which are predicted (see forecasting studies) but still not achieved.

The funding should be sensitive to the needs of SMEs, including the anticipated range of spin-off companies that will emerge. All types of funding should provide mechanisms for encouraging and facilitating potentially complex multidisciplinary research projects, ranging from chemical synthesis to device manufacture, including testing and validation of products.

5.4.2. Standardisation of process, industrial manufacturing, scaling-up

Many companies have already achieved to produce demonstrators (see prototypes of EC co-funded projects) and are prepared to go to small pilot-plant scale but without firm evidence of a strong market pull, companies are not willing to proceed to the next step, the scaling-up.

However, the situation is different if we have a look to component parts, such as electro-conductive yarns. They are already in the market for some years and their development follows a traditional path. As new electro-conductive materials become available, fibre manufacturers find ways of converting them into yarns and fibres. There are several competitors in this market.

On the other hand, the more sophisticated products are either still at research and development stage or are only being made by a small number of companies and the competition is weak. Finished products are expensive and mainly address the entertainment and healthcare market.

However, this barrier is directly linked to costs. Only the standardisation of processes and the development of adequate industrial manufacturing processes will decrease the current high production costs of smart textiles.

A big technological challenge is the scalability of smart textile production. While large-scale production makes better economic sense, this is likely to be a complex task, especially in manufacturing smart textiles with fully integrated systems. Manufacturing standards for smart textiles and components are yet to evolve.

The development of broader scale of smart textiles is certainly a trend within the next decade.

5.4.3. Funding and availability of resources

High cost of research and marketing and a shortage of research funding slow down the development of smart textile products. Research efforts in Europe tends to be driven mainly by governments and the EC. Companies find it very difficult to raise sufficient capital to finance both product research and market development.

Furthermore, there is a need for the support and funding of technology transfer from a research status into products. There is a good state of basic research and on the other side several application-driven projects can be supported by government through respective funding programs. The crossover between these two poles is rarely filled and there is a gap prior to the production transfer. This means a huge financial risk for SMEs, which often cannot be afforded by them.

5.4.4. Partnerships and alliances, collaboration

Current and emerging technologies for the development of smart textiles are specifically adapted for use in the manufacture of textile products. Interdisciplinary collaboration is an essential feature to enable product development – the electronics industry is learning about textile production and fabric architecture; clothing manufacturing techniques are applied to products to specifically solve problems in wearable electronics; and standard commercial textile production methods from the mature textile manufacturing sector are adapted to create smart textiles at mass-produced affordable cost.

The research and development of smart textiles is multidisciplinary and multisectorial.

Stakeholders are convinced that no sector or technology can do it alone. The integration of electro-conductive fibres and yarns, microelectronic components, such as chip modules, user interfaces, such as sensors and displays, power sources and embedded software makes research and development extremely challenging.

In addition, as mentioned earlier, smart textile products should fulfil user needs and expectations in terms of user-friendliness/functionality, costs, fabric resistance, comfort, robustness, reliability and accurate performance. Today's challenge is to fuse research and consumer insights on smart textiles and create multidisciplinary teams of product and electronic designers and engineers. The cluster deals with integration of technologies and functionalities, systems and several applications, e.g. health, well-being, protection and disaster management [47].

There is still a lack of knowledge transfer from the scientific community to companies in the product development phase and a need for tailored interfaces to translate the research results in usable information corresponding to the company needs.

Naturally, there is a difference with regard to cooperation between SMEs and big companies. For SMEs, cooperation with other companies is more important than for large companies. In turn, for large companies cooperation with research organisations is a priority. This implies that big companies are outriders in applying new technologies to extend their product range and find new products. SMEs will first wait and see, if the product launched in the market by a large company has been successful, then they will follow. However, this is not entirely true for the textile industry: there we have a lot of small innovative companies. However, in general, smaller companies do not have the financial capabilities to try out. It is a strategic decision.

Further, there has been a lack of understanding between fashion and electronic industries, particularly with respect to their very different approaches such as product development and planning horizons. Normally, it has been the electronic companies that have taken the lead, often resulting in poorly directed product development that often failed to reflect either true market requirements or the problems of the garment technology integration.

5.5. Drivers

Versatile barriers can be pointed out for the development of smart textiles. However, there are various drivers for the advances in smart textile development and production to enumerate. They can also be clustered into technological, societal and economical as well as strategic nature. In Figure 44 an overview of these drivers is given.

The inherent circumspection of the industry may prove a barrier to progress in the early stages of any new development, since material supply may be limited to research quantities and many companies may prove reluctant to embark on costly scale-up operations without firm evidence of a secure long-term market. There is a strongly held view that unless smart textiles can demonstrate their potential for high added value (meaning that the end-user is

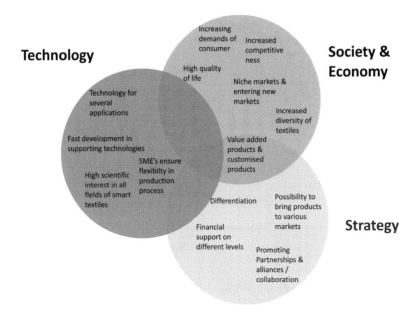

Figure 44.   Drivers for the development and production of intelligent textile products.

prepared to pay a high price), the impetus for their production will be missing. In many cases, the commercial driver is not the ability to do something which is currently not possible but the ability to do the possible more cheaply. On this score, smart materials are likely to be at a disadvantage. In contrast to this relatively negative picture, however, the commercialisation of new smart materials could conversely be catalysed by the growth in a range of start-up and spin-off companies, perhaps triggered by research successes of the academic sector. Whatever be the eventual outcome, there is no doubt that a healthy research and development environment will be an essential prelude for smart textile development over the next five years. *Increased competitiveness* is one driver that can be pointed out when talking about societal and economical drivers. Competitive pressures are already significant for several organisations involved in smart textiles, for instance the competition between research organisations when it comes to available funding schemes for projects.

Increasingly sophisticated electro-conductive fabrics are being developed within the textile sector, although it is not clear how complicated the actual weaving and knitting of electro-conductive yarns is and, therefore, the degree of protection from rivals with low wage costs remain vague. In general, there is little expertise required for weaving electro-conductive fibre meshes once the right materials and constructions have been selected.

The textile sector in its broadest sense could benefit from the development of textile-based components, but much of the value will continue to accrue to the suppliers of the electronic plug-ins, at least for the upcoming next 10 years.

A constant *increase in consumers demands* and a *desire for higher quality of life* account for the rapid growth of smart textile prototypes.

Clothing has always been used to protect the wearer from environment. In the present scenario the protection of the wearer under different circumstances is gaining more and more importance, for instance the concern over terrorist attacks is rising. However, not only this scenario requires a high level of specific protection but also the protection from health and safety risks arises especially from activities in hazardous environments or dangerous

Figure 45. Working principle of the Reach out hats [84]. Reprinted from http://www.tii.se/reform/
projects/itextile/reach.html, with permission of Interactive Institute.

situations, which professionals and emergency services are facing. Looking further in
the professional environment, people working in hospitals also need to be provided with
protection from bacterial contamination.

In everyday situations the concerns over home, children and teenagers' security have
also increased, which give rise to the development and production of smart textiles to
provide *value-added products* to the market.

Next to the safety issue, smart textiles are important for individualisation. The emer-
gence of "mixed" societies has led to diversity, mobility and choice of personal lifestyles.
Therefore smart textiles can be used to enhance the human-to-human interaction, as has
been demonstrated by the Interactive Design Institute in Sweden [84]. They coined the
project "Reach" to investigate the potential for communication and expression to be incorporated dynamically and interactively into the things we are wearing every day (Figure 45).
Within the scope of this project a pair of hats was developed to share experiences in public
space. As the persons wearing these hats approach each other, the textile pattern on the hats
changes and the proximity to each other is expressed (Figure 46).

These trends show that the *diversity of textiles* has increased by the emerging field of
smart textiles. The functions of smart textiles, keeping in view the protection, increase in
performance and applications. As traditional textiles are only meant to protect against cold,
smart textiles will be used to protect against radiation by offering shielding properties, to
alert when there are contaminants in the air or when the body status is critical.

For the moment, a lot of research is dedicated to textiles with sensing applications.
When taking a look at existing prototypes, it gives rise to trust the importance of smart
textiles in the future in our daily lives. The *scientific interest* in the field of smart textiles is
tremendously high and increases everyday. With the announcement of intelligent protective
clothing as one of the lead market initiates of the EC, this trend is fundamentally strengthen
and the importance of smart textiles cannot be denied.

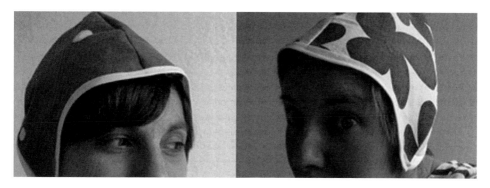

Figure 46. Hats with dynamic patterns [84]. Reprinted from http://www.tii.se/reform/projects/itextile/reach.html, with permission of Interactive Institute.

Certainly, the *commercialisation* of smart textiles can be accounted to strategic drivers. The commercialisation of smart textiles can be catalysed by the growth in start-up and spin-off companies, perhaps triggered by research successes from the academic sector.

Smart fibres, such as those capable of emitting light [85] or possessing switching functions [86–88], also offer obvious commercial opportunities but the fibre manufacturing industry has not adapted to it so far. The development of products with advanced functions, such as piezoelectric textiles, is still in its infancy and far away from commercialisation. In reality these smart fibres are likely to be of utmost interest to high tech industries, such as aerospace for the moment, but the indigenous supply industry is tiny. *Differentiation* and *improved performance* offer an obvious competitive edge.

The smartness in textiles is often gained by chemically modifying the textile surface with "active" inorganic materials that are already available commercially, for instance with a metal coating. Thus, it can be stated that the growth of smart textile materials need not necessarily depend on the development of new materials, such as chemicals, but simply on the evolution of *new uses for the existing materials*. Materials are most likely inorganic compounds, which demonstrate ferroelectric (e.g. piezoelectric), electrostrictive, electro-rheological or particular magnetic properties. In these cases, the supply industry may grow naturally as new applications are realised for existing materials, irrespective of whether or not the materials are recognised as exhibiting smartness.

6. Specific market studies

This section on the roadmap report deals with a specific market study. In the scope of the study we selected five different products (Table 3) and asked companies or organisations involved in the products' development to range the number of synthesized barriers and drivers for their product to enter the market (Table 4), as well as components that are of importance (Table 5).

The products were carefully selected to exemplarily represent the market and to capture the market they stand for. Later, we solely talk about the market they represent in general. An overview of the selected markets, the products and the consultant companies/organisations can be found in Table 3. It needs to be noted that most of the products are still at the prototyping stage. A brief description of each product is given in the accordant market section.

Table 3. Addressed markets and companies or institutions representing the market for the specific market study.

Market addressed	Represented by	Evaluated by
Healthcare	Medical shirt	Smartex
Interior textiles	Carpet	Centexbel
Automobile	Car seat	TITV Greiz
Protective clothing	Firefighters suit	Dalsgaard
Communication and entertainment	Optical fibre display	Brochier

6.1. Healthcare market

The European Commission has stated that the demographic trends in the EU will lead to a quantitative increase in the demand for healthcare. As it is forecasted, the share of population over 65 years of age will increase from 16.1% in 2000 to 27.5% in 2050, hence member states need to pursue accessibility, quality and sustainability of the healthcare system [89]. However, the change in demographics is not the only driving force for a change in the healthcare system. Other trends can also be appointed, which include the following:

- Mandatory search for cost containment.
- Decentralisation of healthcare delivery.

Table 4. Synthesised drivers and barriers for the specific market study.

Drivers	Barriers
Technology	
Technology for several application	Cost
Fast development in supporting technologies	Processing/compatibility with existing product equipment
High scientific interest in all fields of smart textiles	Interconnection of components
SME's ensure flexibility in production process	Education and knowledge of staff
	Standards and quality system
Society & Economy	
High quality of life in terms of comfort, mobility and security	Safety and health constraints
Meeting increasing demands of the consumer	Ease of use, operating & handling
Increased competitiveness	Characteristics of market
Niche markets and entering new markets	Awareness and acceptance of the product
Value-added products and customised products	Promotion of the product
Increased diversity of textiles	Product life cycle/lifespan of product on the market
Strategy	
Differentiation	Service system
Possibility to bring products to various markets	Funding and availability of finances
Financial support on different levels	Standardisation of process, industrial manufacturing, scaling up
Promoting partnerships and alliances	Partnerships and alliances
	Legislation

Table 5.   Components of a smart textile system.

Components	Component material
Sensors	Conductive materials
	Optical fibres
Actuators	Mechanical, e.g. shape memory materials, stimuli sensitive materials
	Chemical, e.g. pH responsive materials
	Thermal, e.g. phase change materials, colour change materials
	Acoustic, e.g. piezoelectric
	Electrical, e.g. colour change materials
Data processing	Conductive materials
	Optical fibres
Powering	Batteries
	Solar energy collectors
	Piezoelectric materials
	Conductive materials, e.g. seebeck materials converting temperature difference into voltage
Communication, e.g. displays, antennas	Conductive materials
	Optical fibres

- Changing disease patterns.
- Impact of information and communication technology including IT support for clinical decision, telemedicine and eHealth.
- More informed and demanding patients.
- Well-being factor and responsibility shift to patients.
- Knowledge management.

These current trends in healthcare and well-being lead to the development of a new market for personal healthcare that can be defined as "products and services to improve the health status and the personal performance outside institutional points-of-care" [2]. Therefore, the most effective means will be more in prevention than in healing and this may happen with a widespread use of vital signs monitoring several user groups, e.g. elderly people and patients with chronic diseases. These tasks may be supported by smart textiles. Hence, different scenarios have already become a reality, including textiles which are equipped with drug-release depots such as microcapsules or garments with integrated sensors to detect biometric parameters. Therefore it is reasonable to think of smart textiles supporting life's quality by the control of motions, vitalisation, preservation of life, monitoring of overweight and body fat, immunisation, control of various biofeedback mechanisms, therapeutical properties or the application of anti-stress therapy. Another challenge will be the reasonable applications against geriatric diseases, especially chronic wounds (Table 6). However, till now garments with biomedical sensing functions have only been available to a limited extent. This might be attributed to the fact that the positioning and the contact precision of sensors are critical for measurements. Above that limitations are still there in terms of resistance to moisture, temperature or mechanical influences. For the time being the signal processing units will be predominantly based on conventional microelectronics, which is a costly factor in the system.

Table 6.   Different sensor principles to monitor vital parameters.

Vital parameters	Measuring principle
Breathing rate	Piëzoresistive fibres, elongation of fibres
ECG/EEG	Electroconductive materials in skin contact sensors
Pulse	Infrared, pressure sensor technology
Stepping frequency	Pressure sensor technology
Sweat	Electrochemical to measure sodium concentration
Temperature	Infrared
Wound healing	Impedance spectroscopy to measure conductivity of blood or plasma

There is a considerable interest in medical sensors and monitors in Europe as part of the European policy on public health, of which the so-called e-Health programme is an important element. There are two key elements in this policy:

Moving from the traditional institution-centred care approach to the patient, citizen-centred approach; a shift from cure to prevention and a growing emphasis on prognosis and early diagnosis.

Involving citizens in the so-called "health paths" as key players, which make them responsible for their own health and to perform relevant tests without involving skilled (and expensive) personnel.

This is a great move towards getting people out of hospitals, shortening their stay and reducing the need for check-ups.

Nonetheless, biomonitoring shirts and bras, such as the Numetrex system by Textronics, have already been marketed.

In Europe considerable research activities are being carried out in this field, mainly driven by the co-funding of the EC. Within the scope of EC co-funded projects a range of prototypes have already been developed, which are for the moment in the validation phase.

In order to assess the healthcare market, focus was put on a medical undergarment measuring different body parameters, such as heart rate, respiration rate, temperature, posture and movement. The system *WEALTHY* (an acronym for "wearable healthcare system") by Smartex, the object evaluated for this market, is equipped with fabric electrodes and connections from electro-conductive steel spun fibres, blended with cotton polyester and woven into a textile structure. The system is powered by an embedded tiny lithium battery [90]. A similar system was developed within the scope of the Proetex project (Figure 47).

6.1.1. Main barriers for smart textiles in the healthcare sector from the experts' perspective

Healthcare applications accommodate an increasing need for security regarding the evaluation of person's well-being and health; this need can be met by relevant products. The potential applications of smart textiles in healthcare are huge and range from recording of vital parameters, the correction of posture to artificial muscles. Although forecasting studies predict a great future for smart textiles in health, the real success stories could not be registered so far. In order to find reasons for this situation, we generated and ranged barriers that affect the development of smart textiles in this sector.

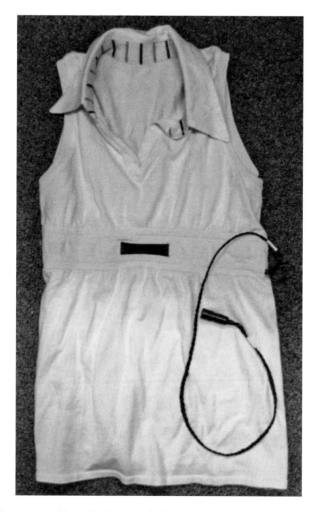

Figure 47. Undergarment shirt with integrated electrodes, developed within the Proetex project. Silver-based yarns were knitted into the garment to shape electrodes. Picture by author.

Following are the main barriers:

- Interconnection of components.
- Standards and quality system.
- Characteristics of the market.
- Funding and availability of finances.
- Standardisation of processes industrial manufacturing, up-scaling.
- Legislation.

It can be seen from Figure 48 that the healthcare market is facing technological, strategic and societal and economic barriers. The maximum barriers from the technological point of view can be attributed to the *interconnection of components* making up the smart textile system as well as the *missing standard and quality system*.

As textiles are worn close to the body, special attention is needed on the techniques applied to connect different components together. Standard techniques used in electronics,

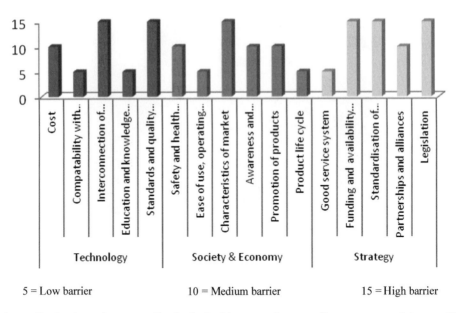

Figure 48. Barriers of smart textiles in the healthcare market according to experts of the specific market study.

such as welding, soldering or gluing, decrease the flexibility of the garment and can lead to discomfort when wearing the shirt. Thus, an appropriate interconnection of components can be achieved by applying textile techniques, such as sewing or embroidery.

The same applies in the case of interconnection materials. Electronic cables and wires lead to discomfort and a decrease the flexibility. For that reason, textile-based solutions, including reliable electro-conductive yarns, are needed.

Another technological barrier that needs to be figured out in the healthcare sector is the *lack of standards* and quality investigations. Due to the young nature of smart textiles, standards to evaluate the performance and compatibility in healthcare are missing and it has to be seen if available medical or electronics standards can be applied. The valida-tion procedures should be carefully selected, long to monitor the long-term performance, and well defined to include clinical trials due to interface between technology and human body. Besides the above-described technological barriers, strategic obstacles are also to be denoted. In this case, the *availability of finances and funding*, the *standardisation of pro-cesses, industrial manufacturing and up-scaling* as well as the *legislation* are encountered as obstacles.

The availability of finances is a barrier in the progress at any stage of new development in smart textiles. There is a great need to ensure that finances are available for research, validation, up-scaling as well as marketing. For example, if the material supply is limited to research quantities, many companies may be disinclined to start costly scale-up operations without firm evidence of a secure long-term market. Finally, from an economic and societal point of view, the *characteristics of the market* represent the most striking barrier. The market is characterised by a high defragmentation, low transparency and low flexibility. These characteristics need to be overcome. Healthcare in general provides a remarkably diverse, high value-added sector for smart textiles, so the promoted products must have the potential for high impact. The further diffusion of technology to healthcare demands for a strong multidisciplinary strategy. Research has hitherto tended to reflect industrial needs

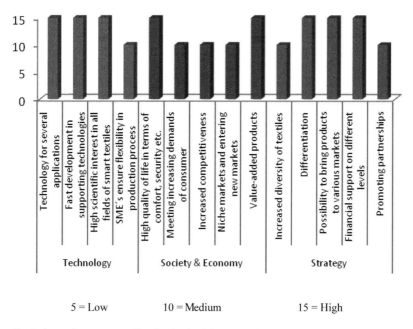

Figure 49. Drivers for smart textiles in the healthcare market according to expert of the specific market study.

The greater transfer of know-how should be promoted by proactive alliances forged from the start of any research initiative rather than towards the end, not only to bring in fundamental expertise from chemistry, electronics and physics but also the active engagement of biomedical researchers. Taking a look at the current funded projects, the consortium ranges from different engineering disciplines of chemistry and physics, but the medical side is mostly forgotten.

6.1.2. Main drivers for smart textiles in the healthcare sector

From the experts point of view the main drivers for smart textiles in the healthcare sector address the technology (Figure 49).

Following are the main drivers:

- Technology for several applications.
- Fast development in supporting technologies.
- High scientific interest in all fields of smart textiles.
- High quality of life in terms of comfort, security etc.
- Value-added products.
- Differentiation.
- Possibility to bring products to various markets.
- Financial support on different levels.

Technologies involved in the development of smart textiles can be used for *several applications*. Thus, major communities that make use of smart textiles may range from hospitals, various geriatric organizations and sleep labs to personal use at home. Especially with the usage of biomonitoring garments at home, a more real-time diagnosis, quicker response times and a more preventative care focus is allowed, which in turn contributes to the lowering of overall healthcare costs. Thus, the key lies in cost reduction on the one hand, and helping to meet people demands on the other hand. Healthcare systems should not only be affordable but also continuously increase the quality of care. The key is to detect diseases much earlier, so that the chances of complete cure are higher and the costs of treatment are lower. Taking a look at the actual situation, there are good tools available to treat acute cases completely and with high quality. The cost situation, however, is suboptimal. If prevention measures are included in the care process and diseases can be detected earlier, targeted treatment could be less expensive and the probability of complete cure could also be much higher. Above that, this approach goes well with the move to enhance disease prevention and consequent reduced pressure on hospitals and doctors, because the monitoring of patients can be continuously done at home.

Further, *high scientific interest* in the field of smart textiles represent a great driver for smart textiles as well as the *fast development in supporting technologies*, such as nanotechnology or microelectronics.

The high scientific interest in the field of smart textiles can be to some extend attributed to the diverse application areas in the healthcare sector. Smart textiles could serve not only as components in diagnostic instrumentation but also in external assistive technologies, limited contact materials, in vivo probes and as implant materials. Smart clothing could serve to provide distributed sensing functionalities to help coordinate a more general gait or movement response, or textiles can serve as a controllable drug delivery vehicle for specific organ and tissue targeting. Here, we might think of smart dressings that react with the patient's or wound's condition by changing colour, or so-called active dressings that in some way react with the condition of the wound to release the healing agent.

At this point, we can directly make the connection with supporting technologies. Nanotechnology and biotechnology, among other technologies, progress well and very fast. This rapid development is of great advantage as these technologies to some extent are used to create smart textiles.

6.2. Automotive industry

In no other sector the demand for smart textiles is so clearly determined by a single powerful industry than in the automotive industry.

The automotive industry is torn between trying to reduce costs on the one hand, and dealing with the high price of performance enhancing technology and environmental compliance on the other hand. This industry can benefit from smart textiles in several domains:

- Seating.
- Ceiling (ambient lighting and integrated loudspeakers).
- Safety belts.
- Airbags.

The automobile sector is a platform that has started to take advantage of smart textiles in different manners.

Smart textiles have already found their way into a number of automotive applications, e.g. in interior parts, including:

- textile switches in the car seat,
- sensors in safety belts,
- sensors in airbags,

- heating in car seats, heating function can be extended to beverage holders etc.
- lighting (Citroën).

An example of textile switches in car seats has been provided by the British company Eleksen. Their Elektex®, a pressure-sensing textile, has found its use in a smart car seat that automatically adjusts itself with the position of the occupant. Sensors in the arm control the motor including the auto recline function that activates when the user occupies the seat and returns to normal position when the seat is unoccupied [62].

Eleksen pressure sensors are also available for use as occupant sensing for airbag deployment. In the United States, for instance, it is already mandatory to have airbag deployment sensors to ensure correct deployment of an airbag in case of collision.

Resistive heating elements in car seats can be increasingly found in the market. The British company Gorix Ltd., for instance, produces woven electronic fabrics that can be heated up on applying a voltage and can be easily incorporated into car seats [91].

In general, automotive applications tend to require high volume and high performance, the ability to function in hostile environments and low cost. Smart textiles are likely to succeed in this sector if they can perform more than one function or can be integrated in a way that reduces assembly and production costs.

A very important mechanical property of textiles is that they reduce weight; they are lighter materials for automotive or airplane applications and thus can reduce energy consumption.

In the scope of this roadmap, a car seat with integrated switches, which was developed by the research institute TITV Greiz in Germany supported by Car Trim [92] was selected to represent smart textiles in the automotive market (Figure 50).

6.2.1. *Barriers for smart textiles in the automotive sector*
Main barriers are as follows:

- Interconnection of components.
- Product life cycle.
- Standardisation of process.

In the automotive sectorm, the *interconnection of components* represents the major barrier in terms of technology for smart textile systems (Figure 51).

The *product life cycle* of smart textiles in the automotive sector is the major societal and economic barrier from the expert's point of view. The question of ecological compatibility and product recycling should play a significant role. If a higher environmental friendliness of textiles, e.g. as heating wires in car seats compared to conventional heating wires could be proven, an important product advantage of textiles when disposing the vehicle can be illustrated.

Further, the most striking strategic barrier is the *standardisation of processes, industrial manufacturing*, and *up-scaling*.

6.2.2. *Drivers for smart textiles in the automotive sector*
Following are the main drivers:

- Technology for several applications.
- High scientific interest in all fields of smart textiles.

Figure 50. Car seat with integrated switches to adjust the position developed by the TITV Greiz supported by Car Trim [92]. Reprinted from http://www.titv-greiz.de, with permission of TITV Greiz.

- Increased competitiveness.
- Niche markets and entering new markets.
- Value-added products.
- Increased diversity of textiles.
- Differentiation.
- Possibility to bring products to various markets.

Advanced electronic systems and high technologies define the vision of a modern car nowadays. Electronics account for around 90% of the innovation in the automotive sector. Smart textiles equipped with electronic functions may also represent an important consumer segment.

The outstanding advantage to the consumer can be appointed to innovative high-quality designs in the automotive interior and new passenger-assistance systems. Textile switches that can be integrated into seats to adjust the seating position, into safety belts to control the radio or into the car ceiling to adjust the inside light represent *one technology* that can be applied *for several applications* (Figure 52). Heatable textiles integrated into car seats can be already found in the market. This technology can be practically implemented into

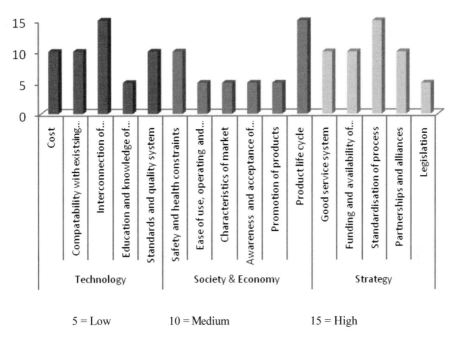

Figure 51. Barriers of smart textiles in the automotive sector according to expert of the specific market study.

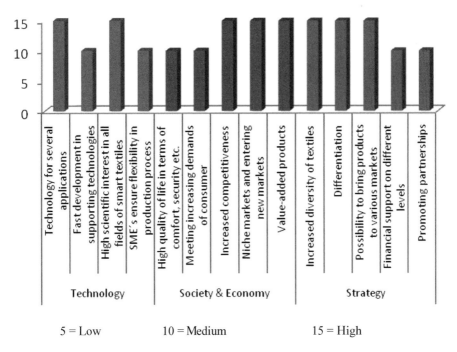

Figure 52. Drivers for smart textiles in the automotive sector according to expert of the specific market study.

the ceiling of a car to allow ambience and distributed heating in the car or it may be used in a beverage holder to keep beverages warm. Above that, there is a *possibility to bring other products into various markets*. Heatable textiles, for instance, would be very suitable for protective clothing as well as sports markets.

A major advantage in the development of smart textiles lies in the fact that there is a *high scientific interest in all fields of smart textiles*. This may significantly influence a fast product development and integration of technologies into automobiles.

As the automotive sector is a highly industrialised sector, companies in this field constantly search for new and advanced technologies and *value-added products* to *increase their competitiveness*. Europe has a strong position in automotive sector and the integration of smart textiles will consolidate that standing in the face of increasing competition. Developments in this area eventually flow down into the mainstream automotive sector as this can be observed with the resistive car seat heating. With the development of smart automotive textiles *niche markets* can be created, which are essential for the survival of textile companies. Thus, the diversity of textiles would increase.

A very important strategic driver is *differentiation*. Differentiation can be achieved in the automotive interior with textiles. Position sensors, for instance, can contribute to better ergonomics of the driver. Light-emitting fibres, woven into the ceiling, enhance the car's ambient lightening.

6.3. Protective clothing

Protective clothing is an issue which is very closely related to the clothing sector. The fields of application are very well defined and the potential solutions and product designs are particularly complex due to their intended purposes. They also need to be harmonised in line with existing regulations, standards and legislations. Protective clothing should protect the wearer against environmental conditions. This protection can be enhanced by capturing the wearer's physiological state and position in the environment. Above that, it may detect dangers in the environment. Thus, a combination of sensor and communication technologies is envisaged.

In order to monitor the vital parameters of a worker, risk data could be sent from the sensor to a central control unit to be in contact with the specialists on site, similar to that proposed for healthcare applications. However, a self-controlling active textile system that, for instance, can react to temperature would be available.

The sensing technology that is needed to monitor the environmental conditions will depend on the use and the operating conditions. It goes without saying that the protection against heat and flames requires different sensors compared to the protection against toxic chemicals in the environment. Under these extreme conditions we also need to distinguish between monitor and alert sensors.

Pioneering work in the field of intelligent protective clothing has been carried out by the Danish company Ohmatex for Viking Life-Saving equipment [93], as depicted in Figure 53. The company has developed a firefighters jacket that alerts the wearer when he is exposed to critical temperatures. The need for the jacket arose because the main cause of death among firefighters in action is heart attack caused by radiant heat. For that purpose thermal sensors were integrated into the jacket to log critical heat levels before they result in physical injury. For that purpose two independent heat sensors are required; a sensor located inside the jacket measures developments in body temperature and a second sensor located on the shoulder alerts firefighters of a rise in critical temperature.

Figure 53. The Viking firefighters suit with integrated sensor and communication technology [93].
Reprinted from http://www.viking-fire.com, with permission of Viking Life-Saving Equipment.

The alerting function is based on visualisation; an array of LEDs is incorporated into
the sleeve of the jacket.

In order to provide sensors that are thermally stable, sensors are encapsulated into
flexible silicon elements in order to withstand temperatures that are in excess of 260°C for
more than five minutes.

The encapsulated sensors are connected to the battery box and displayed by an especially
developed electro-conductive ribbon woven from standard Nomex® yarn spun together with
electro-conductive metal filaments [93].

The Viking firefighter suit is a product that was used to represent the protective clothing
market in this roadmap.

6.3.1. Barriers for smart textiles in the protective clothing sector

The *interconnection of different components* making up the system represents the greatest
technological barrier for protective clothing (Figure 54). The system of protective clothing
is comprehensive as sensors need to be connected with an energy source and a processing
unit that in turn is connected to a communication tool. As the technology is not yet mature
and most of the textile parts are combined with rigid electronic components, the interconnection and the technology to be applied are challenging. Not only the interconnection
is problematic but also the mounting of necessary sensor systems, the energy supply and
the external data transfer in cases where the protective clothing is connected to a local
communication infrastructure and the localisation via GPS or other methods are equally
problematic and actually address all the problems that are for the moment associated with
smart textile systems. Especially in the protective clothing sector, a lot of *standards* are
to be fulfilled and an *extended quality control and testing* is necessary for the garments.
As textiles are now equipped with electronics, new standards and test procedure are to
be defined and it is a challenging task to meet the high quality standards that need to be
fulfilled for protective clothing.

The production process is to be standardised for the high level of quality system.

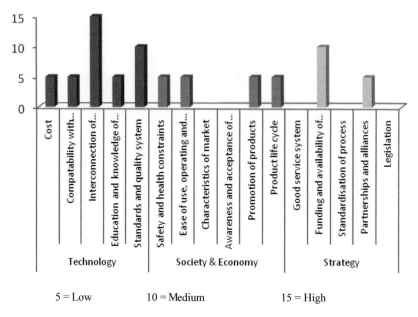

Figure 54. Barriers of smart textiles in the protective clothing sector according to expert of the specific market study.

6.3.2. *Drivers for smart textiles in the protective clothing sector*

Following are the main drivers:

- High scientific interest in all fields of smart textiles.
- Meeting increasing demands of consumers.
- Niche markets and entering new markets.
- Value-added products.
- Differentiation.

From the expert's point of view, there are various drivers for the smart textile systems for protective clothing (Figure 55). First, the *high scientific interest in all fields related to smart textiles* is a driver for the development of intelligent protective clothing. This may be derived from the fact that an intelligent protective garment will preferably compile all six components that make up a smart textile system.

The creation of *value-added products* can be directly correlated to the fulfilment of the *increased consumer demands* and *differentiation*. The innovation and the added value of a firefighter jacket equipped with sensors, communication tools and actuators can be related to the rescue of humans.

Due to the complexity of intelligent protective clothing, *partnerships and alliances* are created between different sectors as well as within the sector. Partnerships are a fruitful tool to accelerate the development of these systems as needed knowledge is transferred much easier.

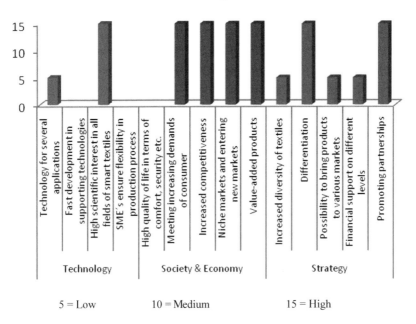

Figure 55. Drivers for smart textiles in the protective clothing sector according to expert of the specific market study.

6.4. Interior textiles (carpet) and construction (building)

The topic of ambient intelligence has received a considerable attention over the past years. It refers to electronic environments that are sensitive and responsive to the presence of people. Electronic devices support people to carry out their everyday activities in an easier way using information and smartness that is hidden in the network connecting these devices. As these devices grow smaller, more connected and integrated into our environment, the technology disappears from our surroundings until only the user interface remains perceivable by users.

Buildings and homes are ideal to integrate components such as sensors, displays, light panels etc. These can be invisibly integrated into wall covers, curtains, floor coverings or furniture. One may think of photonic woven textiles that are displays and wall coverings at the same time, heatable carpets or sofas or light-emitting curtains and floor coverings.

Specific applications may be categorised as "positively beneficial", "useful" or just "nice to have". For example, the use of smart textiles to provide thermal control, including air conditioning (heating and cooling) and the control of solar radiation by the use of curtains would be beneficial in a variety of domestic, office and communal space environments (Figure 56).

Further developments using smart materials such as photo- or thermochromics may offer added value because these materials offer the ability to control light, temperature and colour, leading to positive benefits both in terms of energy efficiency and in the areas of fashion and mood control. To draw an example, International Fashion Machines developed an electro-textile wall panel by using the so-called Electric Plaid™, a woven fabric that exploits reflective colouring (Figure 57). This fabric contains interwoven stainless steel yarns, which are connected to drive electronics. The fabrics are printed with thermochromic inks. Thus the fabric can be programmed to change colour in response to heat derived from the electro-conductive yarns [72].

Figure 56. The Energy Curtain; a project carried out at the Interactive Design Institute in Sweden. The curtain absorbs sun energy during the day and releases it in the form of light in the evening [7]. Reprinted from http://www.tii.se, with permission of Interactive Institute.

In Europe research is on around carpets that react upon pressure in order to detect intrusion or obstruction, or carpets that light up in response to heat. Although the textile content is high in carpets, the smartness does not necessarily need to be textile-based because a carpet is just put on the floor and typical textile-related properties, such as drapability, are not of primary importance.

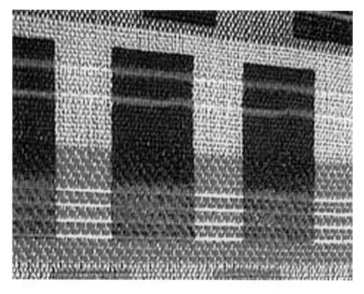

Figure 57. Electric Plaid™, colour-changing textile wall panels [72]. Reprinted from http://www.ifmachines.com/, with permission of Maggie Orth, IF Machines.

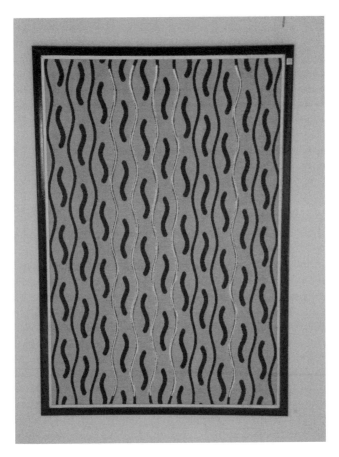

Figure 58. Intelligent (luminous) carpet equipped with sensors and light-emitting fibres developed at Centexbel; the carpet is exhibited presently at Living Tomorrow in Vilvoorde, for the purpose of surveillance of elderly people in an apartment [94]. Reprinted from http://www.centexbel.be, with permission of Centexbel.

A carpet equipped with pressure sensors and light-emitting fibres developed by the Belgian-based Centexbel was chosen to represent the interior textiles market in this roadmap.

The floor carpet in Figure 58 detects the presence, location and displacement of people on the carpet. It can switch on the light when someone gets out of bed. If someone lies on the carpet for a defined time period, a remote emergency service gets alerted. The luminescent "waves" are turned on and off depending on the location and the displacement of the person on the carpet [94].

6.4.1. Barriers for smart textiles in the construction sector/interior textile sector

The main barriers are as follows:

- Interconnection of components.
- Funding and availability of finances.
- Standardisation of process.

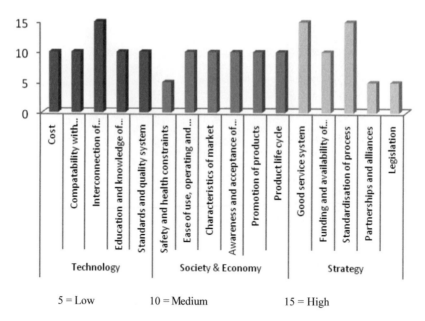

Figure 59. Barriers of smart textiles in the interior textiles sector according to expert of the specific market study.

As far the markets described above, the *interconnection of components* can be regarded as the greatest challenge for intelligent interior textiles (Figure 59). Further, the experts have pointed out that presently the *service system* to maintain smart interior textiles is insufficient. In order to raise the awareness of consumers about such a carpet, it needs to be convincingly demonstrated at the sale point. The increased security and safety level and the unproblematic care properties are to be clearly pointed out. The lacking *industrial manufacturing* also represents a striking barrier for the time being.

6.4.2. Drivers for smart textiles in the construction sector/interior textile sector

The man drivers are as follows:

- Technology for several applications.
- Fast development in supporting technologies.
- High scientific interest in all fields of smart textiles.
- High quality of life in terms of comfort, security etc.
- Niche markets and entering new markets.
- Increased diversity of textiles.
- Possibility to bring product to various markets.
- Financial support on different levels.
- Promoting partnerships.

According to experts, the drivers for smart textiles used in interior textiles are numerous (Figure 60). From a technological point of view, it can be pointed out that the *technology can be used for several applications*, a fast development in supporting technologies can be documented and there is a high scientific interest in all fields of smart textiles.

From a societal and economic point of view, the drivers such as *higher quality of life* in terms of comfort, mobility and security for the consumer, the *increasing demands of*

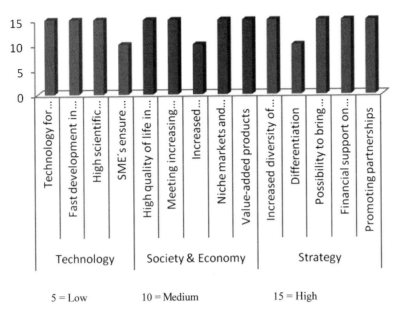

Figure 60. Drivers for smart textiles in the interior textiles sector according to expert of the specific market study.

consumers in terms of comfort and safety can be met and *value-added products* can be offered in the market that may secure the European textile market. Above that, the *diversity of textiles* will be increased and *niche markets* can be created.

A strategic driver for the implementation of smart textiles in interior textiles is the *possibility to bring products to various markets*. Pressure sensors in carpets, for instance, will be of importance in homes, but the same technology can also be integrated in car seats for instance as pointed out in a previous chapter. Further, the *funding possibilities* that are offered on regional, national and European level emphasize the importance of smart textiles. The European Commission created lead market initiatives, among which are intelligent protective textiles [95]. Smart textiles imply a multidisciplinary approach and thus, *partnerships and alliances are promoted*.

6.5. Communication and entertainment

The fields of communication and entertainment can be regarded as being new from a textile point of view. The example of a jacket with an integrated mp3 player is already market ready and several electronic companies (e.g. Infineon, Philips) and various consumer brands count this product to their product range (e.g. Rosner, Levi's,O'Neill, Hugo Boss). However, this new type of product will still need to proof, if such solutions are accepted by the user or if electronic devices which are already extremely mobile and easy to operate win the competition.

Clip-on products, such as mp3 players and phones led the way for the wearable electronic entertainment market and thus the market is more diverse. These are "aspirational" products for which functionality is less important than brand and design content. These applications are easy to understand for the public and early adopters are willing to take up. However, it is difficult to judge, as just pointed out, whether the market will develop beyond the take-up by early adopters.

Figure 61. The Philips Lumalive couch exemplifies how Lumalive light-emitting textiles can be used in the home, or professional environments like lobbies and offices [39]. Reprinted from http://www.lumalive.com/AboutUs/Press, with permission of Philips Lumalive.

In the field of communication niche markets are evolving which were 10 years ago entirely closed for the textile industry. Smart textiles which ensure portable communication technology by means of textile user interfaces, such as displays are market ready. One example is Philips Lumalive [39].

The Lumalive displays are made up by tiny LEDs (Light Emitting Diodes) which are integrated into textiles. The application is not limited to clothing, like shirts or jackets; it can also be integrated into home furnishing and floor coverings (Figure 61).

In general, displays represent a large part of the total cost and power consumption of most appliances. In turn, for the personal environment, low power and lightweight displays are important. Also, multi-modal interfaces that include gesture recognition and input from sensors to support speech and sound driven systems are crucial for people moving around. Personal interface management systems need to be considered as well to enable personal devices to interact with each other and to use the interface devices available in different environments, such as head-up displays in cars or office wall.

An optical fibre display developed by France Telecom and woven by Brochier was the evaluated product in the market study. Optical fibres were used to create textile-based displays, so-called Optical fibre flexible display (OFFD). For this purpose optical fibres as

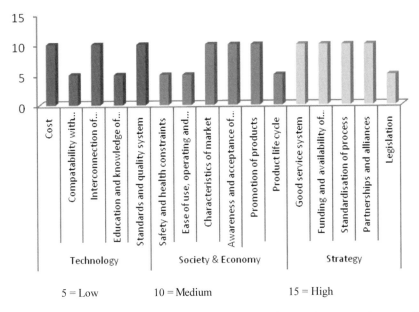

Figure 62. Barriers of smart textiles in the communication and entertainment market according to expert of the specific market study.

well as conventional fibres make up a woven fabric structure. A small electronic device, integrated into the textile system, controls the LEDs that illuminate groups of fibres. Each group provides light to one pixel (given area) on the matrix so that various patterns can be displayed in both, dynamic and static manner. The fact that it possesses a very thin size and an ultra light weight makes the structure interesting [96].

6.5.1. Barriers for smart textiles in the entertainment/communication sector

From the expert's point of view there are no significant barriers for optical fibre displays integrated into textiles (Figure 62).

However, the optical fibre display has not yet reached the stage of market maturity. This may be explained by the fact that complex display technologies, even if these serve exclusively to transmit signals, interfere with the wear and care properties of conventional textiles due to their construction.

6.5.2. Drivers for smart textiles in the entertainment/communication sector

Main drivers are as follows:

- Fast development in supporting technologies
- High scientific interest in all fields of smart textiles
- SME's ensure flexibility in production process
- Niche markets and entering new markets
- Value-added products
- Possibility to bring products to various markets
- Promoting partnerships

For the communication and entertainment sector the drivers for smart textiles are more or less the same as for the other markets (Figure 63). The expert nominated the

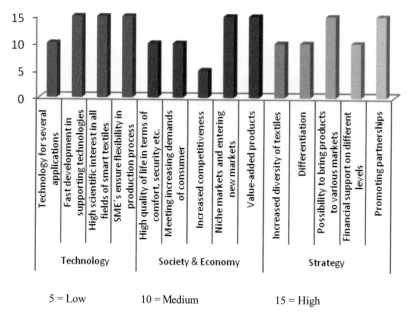

Figure 63. Drivers for smart textiles in the communication and entertainment market according to expert of the specific market study.

fast development in supporting technologies, the *high scientific interest in all fields of textiles* and the fact that *SMEs ensure flexibility in production process* as technological drivers.

The creation of niche markets and the possibility to enter new markets as well as the creation of value-added products and customised products account for societal and economic drivers.

The possibility to *bring products to various markets* and the *promotion of partnerships and alliances* attribute strategically to the development of smart textiles in the entertainment and communication market.

6.6. Comparison of addressed markets

Smart textile systems will improve and extend the functionality of textile products. This in turn will have an effect on traditional (Figure 64) and new fields (Figure 65) of application for textiles. In principle, the fields of identified applications can also be regarded as markets that need to be addressed [98].

The five selected markets can be regarded as particularly interesting in terms of developing niche markets. Areas which demonstrate very attractive perspectives are the markets of smart protective clothing and smart healthcare textiles.

In protective clothing, the integration of sensors will significantly improve the protective function. Communication technologies, such as displays or wireless tools, will enable the wearers or the persons monitoring scenarios to be alerted at the right moment, thus establishing a safe and effective system.

In healthcare applications, the challenging concept of mobile and independent health monitoring is addressed. The integration of sensors in textile systems will enable intelligent biomedical clothing systems. Above that, powering and wireless communication tools are

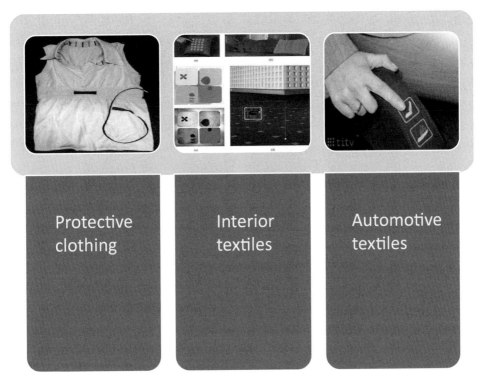

Figure 64.   Smart textiles for "traditional" fields of application. Examples from left to right: Prototype of an undergarment for a fire-fighter, developed within the Proetex project [55]. Picture by author; interactive pillow by Interactive Design Institute Göteborg [7]. Reprinted from http://www.tii.se, with permission from Interactive Institute; integrated switch into a car seat developed by TITV Greiz [92]. Reprinted from http://www.titv-greiz.de, with permission of TITV Greiz.

to be adapted and preferably be textile-based to ensure flexibility and a good integration into the whole textile system.

6.7. *Major components in smart textile systems*

In the scope of the market study, the experts rated the importance of major components making up a smart textile system (Table 7). The communication function is of great importance for all selected application fields. For that, the development of appropriate electro-conductive materials is needed. Powering and sensors are also components of importance. The research and development activities are vivid around sensors, thanks to the Smart Fabrics-Interactive Textile and Flexible Wearable Systems (SFIT) funding cluster of the EC as described above. The powering of a smart system, which is of utmost importance, still represents the greatest challenge. Research is already being conducted into the field of integration of flexible solar cells and flexible batteries, but the textile character is still insufficient. It gets apparent that suitable solutions need to be found and investigated. Actuators for the moment are of minor importance. This might be attributed to the fact that actuators are very sophisticated elements and that presently it is of more relevance to achieve reliable and stable sensors, power sources and communication tools with good performance.

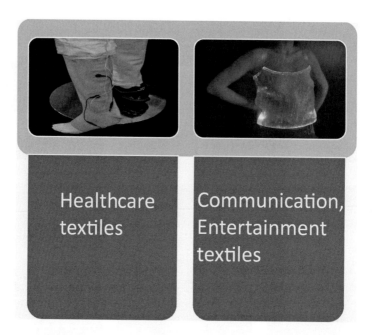

Figure 65. Smart textiles for "new" fields of application. Left example: electrotherapeutic sock developed by Ghent University within the Lidwine project; right example: light-emitting shirt by Luminex [99]. Reprinted from http://www.luminex.it, with permission of Luminex. The photograph is the property of Luminex S.p.A. The product in the photograph is made from LUMINEX® FABRICS.

6.7.1. Barriers according to different markets

In general, there are fewer barriers rated as high by the experts compared to drivers. This indicates that the potential for smart textile application is great and they represent a great opportunity for the addressed market.

When opposing the major barriers to the selected markets as shown in Table 8, it can be seen that the interconnection of components is the greatest challenge for the moment that needs to be faced in the development of smart textile systems for all markets. The

Table 7.   Major components of a smart textile system versus the selected five markets.

Market	Sensors	Actuators	Data processing	Powering	Communication
Healthcare	Conductive materials and optical fibres		Conductive materials and optical fibres	Batteries	Conductive materials and optical fibres
Automotive	Conductive materials	Chemical and Thermal			Conductive materials
Protective clothing	Conductive materials			Batteries	Conductive materials
Interior textiles	Conductive materials		Conductive materials	Piëzoelectricm aterials	Conductive materials
Communication			Conductive materials	Batteries	Conductive materials and optical fibres

Table 8.   Major barriers of smart textiles for the five selected markets.

Market	Interconnection of components	Standards and quality system	Characteristics of market	Product life cycle	Good service system	Funding and availability of finances	Standardisation of processes	Legislation
Healthcare	■	■	■			■	■	■
Automotive	■			■			■	
Protective clothing	■							
Interior textiles	■				■		■	
Communication								
	Technology		Society and economy		Strategy			

standardisation of the process, the industrial manufacturing and scaling-up follows, as these are the issues of great significance in the healthcare, automotive and interior textile sectors.

Surprisingly, the costs involved in the development and production of smart textiles are not rated as a very significant barrier by any market. This gives rise to the belief that the added value and opportunities that these systems involve are much more important.

In the scope of the Delphi study, a number of barriers were synthesised and the industrial experts answering the Delphi questionnaire were asked to rate their importance for selected applications (Figure 66). These barriers and application fields are in accordance with the

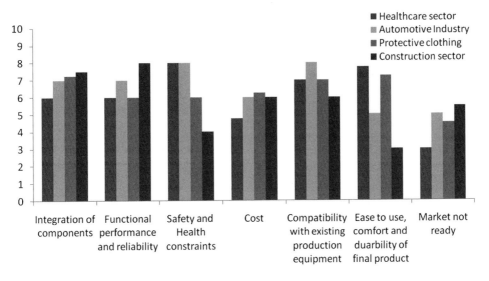

Figure 66.   Barriers of smart textiles according to Delphi study.

A. Schwarz et al.

Table 9. Major technological drivers for smart textiles for the five selected markets.

Market	Technology for several applications	Fast development in supporting technologies	High scientific interest in all fields of smart textiles	SME's ensure flexibility in production process
Healthcare	■	■	■	
Automotive	■		■	
Protective clothing			■	
Interior textiles	■	■		
Communication		■	■	■

selected market study. The integration of components as well as the compatibility with existing production equipment and safety and health constraints are rated as major barriers.

6.7.2. *Drivers according to different markets*

The experts rated a number of drivers as very important. In the following tables (Tables 9–11), the major technological, societal and economical as well as strategic drivers for the selected markets are presented.

In the previous paragraph it was stated that costs are not a major barrier. Instead, cost can be actually seen as a driver: once the prototyping phase is successfully finished, the cost of production can be cut by higher production numbers. In the end, smart textile systems will contribute to save costs.

Table 10. Major societal and economic drivers of smart textiles for the five selected markets.

Market	High quality of life	Meeting increasing demands of consumers	Increased competitiveness	Niche markets and entering new markets	Value-added products	Increased diversity of textiles
Healthcare	■				■	
Automotive			■	■	■	
Protective clothing		■	■			
Interior textiles	■			■		■
Communication				■	■	

Table 11.   Major strategic drivers of smart textiles for the five selected markets.

Market	Differentiation, increasing number of product types and variants	Possibility to bring product to various markets	Financial support on different levels (regional, national, EU)	Promoting partnerships and alliances
Healthcare	■	■	■	
Automotive	■	■		
Protective clothing	■			■
Interior textiles		■	■	■
Communication		■	■	■

During the Delphi study, the industrial experts were also asked to rate the drivers for smart textiles, the result can be seen in Figure 67. The creation of value-added products is the most striking driver according to this study.

7. Overall recommendations – what do we do next?

A number of common subjects have emerged from the market-specific studies in the previous sections of this roadmap. Prime requirements for the future of smart textile systems are as follows:

- A coherent technical strategy.
- Multidisciplinary approach.
- Market application focus.

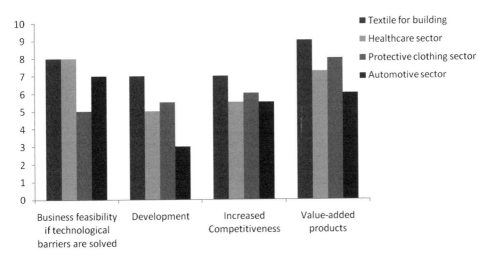

Figure 67. Drivers for smart textiles according to Delphi study (1 = low; 10 = high).

In the development process, it is essential that research initiatives actively include the industry, especially SMEs, as the industry is the direction-setting community of research and development activities. Thus, integrated industry-driven research approaches should be favoured.

A major recommendation of this study is that the technical focus should be based around the following issues:

- Integration of components making up a smart textile system.
- Powering solution for smart textile systems.
- Performance validation of smart textile systems to overcome the technically related weakness and to evaluate the compatibility with the human body (clinical tests).
- Further development work is required, which documents the performance and safety of new types of smart textile systems beyond all doubts by means of new test procedures.
- Product manufacturing and industrial upscaling of the production process.
- This reflects the view of the Delphi study and the specific market study, as these issues are the major barriers to the exploitation of smart textile systems.
- It is in the targeted removal of today's application bottlenecks, in the practical implementation of smart textile technologies and in the subsequent added values where the EU member states can best derive a commercial advantage.
- The technical focus can be achieved by a stronger focus on funding schemes that also reduce the application rate.

Another major recommendation is that the interlinking between technologies and applications should be strengthened. This can be achieved by enhancing spin-off creation.

For the application of smart textile systems, new standards for the manufacturing of these products and for the performance testing need to be developed. A recommendation is to create a body to evolve the necessary standards, to oversee them and to accept the overall responsibility.

A further major recommendation in order to embrace the multidisciplinary approach in the field of smart textiles is to adapt education accordingly by focussing on multidisciplinary education.

Most important, a higher interaction between industry and academia facilitating an effective technology transfer is needed for the growth and prosperity of European smart textiles.

Knowledge created at different education centres and research institutes must be harvested by the textile sector. Creation of national or European fellowships for a time-restricted (e.g. 1 year) stay at such laboratories (textile students, post-graduates going to chemical institutes, physical labs etc.) may overcome this barrier.

Not only the technology transfer between academia or research organisation and industry needs to be strengthened but higher education should also be adapted. As the world is changing at a faster pace and achievements in science and technology are becoming more and more impressive, it is very important to clearly identify the education and training that should be encompassed in the field of applied science.

For this reason courses should be intensified, which should include areas of science such as electronics, polymer science, chemistry, physics and biotechnology at graduate-level education.

The creation of partnerships is closely related to the issue of education. There is a great need for the creation of multidisciplinary centres with advanced knowledge and own pilot production facilities that are essential for supporting the industry in taking its products to the final market. To overcome the gap between research and production, it is important to develop an in-depth knowledge about the sector and potential partners in order to better organise knowledge-based trans-sectorial networking.

This can be supported by the following actions:

- Support inter-community dialogue via workshops.
- Managers and engineers should receive training to raise their awareness on the possibilities for trans-sectorial cooperation and managing organisational changes [100].
- Mapping of potential partner sectors for textiles, focussing on time horizons, product and process development phases and sectorial quality standards.
- Better mapping of existing knowledge to allow knowledge exchange. One pioneering initiative is the EC co-funded SYSTEX project [101].
- Promote existing technologies to companies and establish collaboration by that. For instance, fabric companies, coaters and chemical companies should be brought together to develop consumer products.
- Encourage the necessary machinery and training investment among companies.

The Foresight Material Panel [83] even went one step further by suggesting the establishment of an European Centre for smart materials. This can be done only by focusing on smart textiles. The building stone for this initiative should be the various current European project initiatives to create a "hub and spoke" network for the fast-moving nature of technical developments and its broad market sectorial appeal, together with a strong management function with a high degree of flexibility.

A focus should also be put on the promotion of smart textiles. Research action should be promoted to get accepted by scientific and medical communities. The developments of electro-conductive polymers/fibres/fabric expertise should be promoted, as they are the building blocks for smart textile systems.

The acceptance of smart textile products by the user can be strengthened by introducing technologies in smaller stages to give the user the opportunity to keep track of the developments and by offering a good service system in the form of user manuals, service hotlines, info spots at the points of sale to guarantee end user education.

Acknowledgements

The authors thank their partners within the European Commission co-funded project Clevertex (N° FP6-2003-NMP-TI3-main) for their assistance to realise this document. The authors express their gratitude to the EC for their financial support. Special thanks are attributed to the external experts who helped realising the specific market study. Hence, thanks to Ohmatex, Smartex, Brochier, TITV Greiz and Centexbel.

References

[1] http://www.euratex.org
[2] A. Lymberis and D. De Rossi, *Wearable eHealth systems for personalised health management*, in *Studies in Health Technology and Informatics*, vol. 108, A. Lymberis and D. De Rossi, eds., IOS Press, Fairfax, VA, 2004, pp. 75–78.
[3] R. Galvin, Science 280 (1998) p. 803.
[4] http://www.tfl.gov.uk/gettingaround/1106.aspx
[5] R.N. Kostoff and R.R. Schaller, IEEE Trans. Eng. Manag. 48(2) (2001) pp. 132–143.
[6] M. Schwartz (ed), *Encyclopaedia of Smart Materials*. (ISBN: 0-471-17780-6). Wiley, New York, 2002.

[7]   www.tii.se

[8]   X. Tao, *Smart technology for textiles and clothing*, in *Smart Fibres, Fabrics and Clothing*, X. Tao, ed., Woodhead, Cambridge, UK, 2001, pp. 1–5.

[9]   S. Vassiliadis, *Automation and the Textile Industry*, M. Beigl, S. Intille, J. Rekimoto and H. Toknda, eds. Educational Institution of Piraeus & Guimaraes Universidade do Minho, Greece, 1996, p. 148.

[10] E.P. Scilingo, F. Lorussi, A. Mazzoldi and D. De Rossi, IEEE Sensors J. 3(2–3) (2003), p. 460–467.

[11] R. Paradiso and D. De Rossi, *Multichannel techniques for motion artifacts removal from electrocardiographic signals*. Proceedings of the 28th IEEE EMBS Annual International Conference, New York City, 2006.

[12] M. Catrysse, R. Puers, C. Hertleer, L. Van Langenhove, H. Van Egmond and D. Matthys, Sensor Actuat. A-Phys. 114(2–3) (2004), p. 302–311.

[13] M. Rothmaier, M. Phi Luong and F. Clemens, Sensors 8 (2008) pp. 4318–4329.

[14] J. Meyer, P. Lukowicz and G. Troester, *Textile pressure sensor for muscle activity and motion detection*. Paper presented at 10th IEEE International Symposium on Wearable Computers, Montreux, Switzerland, October 11–14, 2006 (pp. 69–72).

[15] M. Sergio, N. Manaresi, M. Nicolini, D. Gennaretti, M. Tartagni and R. Guerrieri, Sensor Lett. 2(2) (2004) pp. 153–160.

[16] F. Carpi and D. De Rossi, *IEEE Trans. Inform. Tech.* 9(3) (2005) pp. 295–318.

[17] J. Whan Cho, J. Won Kim and N. Seo Goo, Macromolecular Rapid Comm. 26 (2005) pp. 412–416.

[18] J.A. Balta, F. Bosia, V. Michaud, G. Dunkel, J. Botsis, and J.A. Månson, *Smart Mater. Structure* 14 (2005), pp. 457–465.

[19] Q. Meng, J. Hu, Y. Zhu, J. Lu and Y. Liu, J. Appl. Polym. Sci. 106(4) (2007) pp. 2515–2523.

[20] I. Liao, A. Wan, E. Yim and K.W. Leong, J Control. Release 104(2) (2005) pp. 347–358.

[21] E.-R. Kenawy, G.L. Bowlin, K. Mansfield, J. Layman, D.G. Simpson, E.H. Sanders, G.E. Wnek, et al., J Controlled Release 8 (2002) 57–64.

[22] J. Zeng, L. Yang, Q. Liang, X. Zhang, H. Guan, X. Xu, et al., J. Control. Release 105 (2005) pp. 43–51.

[23] http://www.dephotex.com

[24] M.B. Schubert and H.J. Werner, Materials Today 9(6) (2006), p. 42–50.

[25] http://www.siliconsolar.com/visual-directory/12-v-consumer-ready-panels.html

[26] http://www.enfucell.com/products-and-technology

[27] http://www.powerpaper.com/home.php

[28] G. Johnson, *Thin battery technology*. Paper presented at the Meeting of the Printed Electronics USA 2006, Phoenix, AZ, USA, 2006.

[29] http://www.bluesparktechnologies.com

[30] H. Nishide, S. Iwasa, Y.J. Pu, T. Suga, K. Nahasara and M. Satoh, Electrochem. Acta. 50 (2004) pp. 827–831.

[31] T. Suga, H. Konishi and H. Nishide, Chem. Comm. 17 (2007) pp. 1730–1732.

[32] http://www.nec.co.jp

[33] T. Starner, IBM Systems J. 35 (1996) pp. 618–629.

[34] http://www.scottevest.com/v3_store/Fleece_Jacket.shtml

[35] C. Hertleer, L. Van Langenhove, and H. Rogier, Adv. Sci. Tech. 60 (2008), p. 64–66.

[36] C. Hertleer, A. Tronquo, H. Rogier and L. Van Langenhove, Text. Res. J. 78(8) (2008) pp. 651–658.

[37] F. Declercq, H. Rogier and C. Hertleer, IEEE Trans. Antennas Propagation 56 (2008) pp. 2548–2554.

[38] http://www.gizmag.com/go/3043/picture/5674/

[39] http://www.lumalive.com/AboutUs/Press

[40] http://www.vorwerk-teppich.de/sc/vorwerk/template/bildmeldung_thinkCarpet_en.html

[41] http://www.textile-wire.ch/downloads/neu_textile_wire_doc_de.pdf

[42] http://www.izm.fraunhofer.de/EN/About/strat_allianzen/uni/HeterogeneousTechnology AllianceHTA.jsp

[43] http://www.ohmatex.dk/images/lederbaand-030.jpg

[44] C. Kallmayer, T. Linz, R. Aschenbrenner and H. Reichl, mstnews, 2 (2005) pp. 42–43.